LIQUID CHROMATOGRAPHY
ESSENTIAL DATA

D. Patel
Department of Medicine & Pharmacology, Section of Molecular Medicine, University of Sheffield, Royal Hallamshire Hospital, Sheffield, UK

JOHN WILEY & SONS
Chichester · New York · Weinheim · Brisbane · Singapore · Toronto

Published in association with BIOS Scientific Publishers Limited

 Published in association with BIOS Scientific Publishers Ltd, 9 Newtec Place, Magdalen Road, Oxford OX4 1RE, UK.

British Library Cataloguing in Publication Data
A catalogue record for this book is available from the British Library.

ISBN 0 471 97270 3

Typeset by Marksbury Multimedia Ltd, Midsomer Norton, Bath, UK
Printed and bound in UK by Page Bros, Norwich, UK

CONTENTS

Tables

PREFACE

It is the aim of this data book to present in a concise manner the core information required for liquid chromatography of both small and large molecules. This book seeks to provide a comprehensive reference source for support matrices, sample preparation, solvent systems, and methods for detection and analysis after chromatography. In addition, an extensive troubleshooting guide is provided to help the reader solve potential difficulties that may be encountered in carrying out liquid chromatographic separations. I have tried to create a balance between basic information and the more specialized areas of liquid chromatography, but in creating such a balance it is inevitable that some areas will receive less attention than they merit and I am conscious of this fact. However, I hope that the reader will find this book to be instructive and useful in their work.

Finally, I am grateful to David Rickwood and those at BIOS for their help during the preparation of this book.

D. Patel

HEALTH HAZARDS AND PRECAUTIONS

Many chemicals commonly used for liquid chromatography are either toxic or flammable whilst the status of others is unknown. The reader must acquaint themselves with the precautions required for handling the chemicals mentioned in this text. Acetonitrile and methanol are both toxic and flammable, ethanol and hexane are flammable, and dimethyl sulfoxide is an irritant. Great care must be taken when handling such reagents.

It is essential to avoid any protease and nuclease contamination and therefore, where possible, all solutions should be autoclaved. Solutions which cannot be autoclaved should be prepared in glass double-distilled, autoclaved water and then filtered through a Millipore filter. In order to avoid other impurities, it is recommended that the solvents used are of high if not HPLC grade purity.

Chapter 1 CHROMATOGRAPHIC SEPARATION MECHANISMS

1 Introduction

Chromatography is one of the most important separation methods used by chemists and biochemists for the separation of both large and small molecules. Chromatography is the differential separation of sample components between a mobile phase (liquid or gas mixture) and a stationary phase (column matrix). Heftman [1] defined chromatography as "a method of analysis in which the flow of solvent or gas promotes the separation of substances by differential migration from a narrow initial zone in a porous sorptive medium". Many different types of chromatography have been developed, including paper, thin layer, gas, and liquid chromatography. Owing to space limitations, this volume of the Essential Data Series will be restricted to liquid chromatography.

Liquid chromatography can essentially be divided into conventional (low pressure) and the now more frequently used high-performance chromatography (*Table 1*). High-performance liquid chromatography (HPLC) is also referred to as high-pressure liquid chromatography, and is distinguished from conventional liquid chromatography by the use of sophisticated, high efficiency and sensitive instrumentation. There are advantages of HPLC over conventional liquid chromatography, and these include increased speed, higher resolution, higher sensitivity, greater reproducibility, and better sample recovery. Many different types or classes of liquid chromatographic techniques have evolved and are employed for both conventional and high-performance chromatography. The more common and widely used types of liquid chromatography include:

1. size exclusion chromatography;
2. hydrophobic interaction chromatography;
3. hydroxylapatite chromatography;
4. ion-exchange chromatography;
5. ion-moderated partition chromatography;
6. normal phase chromatography;
7. reversed-phase chromatography; and
8. affinity and chiral chromatography.

2 Size exclusion chromatography

Size exclusion chromatography is also known as gel permeation chromatography, gel filtration chromatography, or molecular sieving. Unlike other liquid chromatographic techniques, which separate molecules according to their surface properties, this technique separates molecules by their size or, more precisely, by the hydrodynamic volume of the molecule. The separation is achieved by repeated diffusion of the sample molecules into and out of the pores of the gel or

functional group of the support matrix. The functional group, or ligand, attached to the support matrix determines the selectivity of the matrix in binding molecules. Sample molecules with very few nonpolar or hydrophobic groups weakly interact with the hydrophobic interaction chromatography column, and tend to elute quickly from the column. This results in shorter retention times (R_t). Conversely, sample molecules that interact strongly with the hydrophobic interaction chromatography column tend to have longer retention times (R_t). Hydrophobic interaction chromatography is commonly used for the analysis or purification of proteins, peptides, and nucleic acids.

4 Hydroxylapatite chromatography

High-performance hydroxylapatite (hydroxyapatite) chromatography separates macromolecules by differential binding to phosphate or calcium sites on the support matrix. The differential surface binding is primarily ionic in nature; however, it also has specific affinity properties. As binding beads that make up the support matrix. The pores have a carefully controlled range of sizes. Molecules with sizes greater than the pore diameter of the support matrix cannot enter the pores, and are excluded and quickly eluted from the column in the void (dead) volume (V_0). Molecules with sizes smaller than the pore diameter enter the pores and differentially elute in volumes greater than the void volume (V_0) and hence at various retention times (R_t). The molecules normally elute in order of decreasing size. Size exclusion chromatography has various applications including the separation of sample components, estimation of molecular weights, determination of molecular weight distributions, determination of equilibrium constants, and desalting.

3 Hydrophobic interaction chromatography

Hydrophobic interaction chromatography (HIC) is a versatile method for the purification and separation of molecules based on differences in their surface hydrophobicity. The separation process occurs by differential hydrophobic interactions of the sample molecules or components with the can be ionic, the charge density and isoelectric point of the molecule may be important factors in the separation process. The affinity-type interaction can be with the calcium or phosphate molecules of the hydroxylapatite support, as is the case with nucleic acids [2]. Hydroxylapatite chromatography is a technique commonly used for the separation and purification of proteins, in particular enzymes and antibodies, and nucleic acids, especially single-stranded from double-stranded DNA.

5 Ion-exchange chromatography

Ion-exchange chromatography is among the most precise methods for the fractionation of biological substances. Ion-exchange chromatography separates molecules based on the ionic charge or isoelectric point (pI) of molecules at a given pH. Separation is achieved by differential ionic interaction of the sample molecules (or ions) with the negatively or positively charged functional groups of the support matrix. The separation proceeds because ions of opposite charge are retained to different extents. Sample ions that interact weakly with the charged support matrix (ion exchanger) will be weakly retained on the column and elute early. Those sample ions that interact strongly with the ion exchanger will be strongly retained on the column and elute at later retention times (R_t). Elution of sample ions from the column occurs by increasing the ionic strength (ionic concentration) of the buffers. In addition to the ion-exchange effect, other types of binding may occur. These effects are small and are mainly due to van der Waals forces and nonpolar interactions. Ion-exchange chromatography is widely used for the analysis, separation, or purification of proteins, peptides, amino acids, and nucleic acids.

6 Ion-moderated partition chromatography

Ion-moderated partition chromatography is a high-performance liquid chromatographic technique which separates molecules based on a number of different chemical and physical characteristics. The separation mechanisms employed are ion exchange, ion exclusion, normal and reversed of steroids, lipids, phospholipids, fatty acids, and other organic compounds.

8 Reversed-phase chromatography

Reversed-phase chromatography separates molecules based on their hydrophobicity. Separation occurs by differential hydrophobic interactions of sample components with the nonpolar, hydrophobic functional groups attached to the matrix. The hydrophobicity of the functional groups on the matrix is proportional to the length of their carbon chains. Sample molecules with very hydrophobic moieties interact strongly with the column, and will tend to elute at later retention times (R_t). Those molecules with polar groups or few hydrophobic moieties will interact weakly with the column and elute quickly. Increasing the polar (aqueous) component of the eluant increases the retention of the solutes. The typical conditions used, low pH and high concentrations of organic solvents, favor denaturation. This technique is used in many circumstances in which denaturation may be phase partition, size exclusion, and ligand exchange. One or more of these mechanisms may be operating at once on the compounds being separated. Ion-moderated partition chromatography is used for separating carbohydrates, organic acids and bases, alcohols, and a variety of metabolites from food, biological and industrial sources.

7 Normal phase chromatography

Normal phase chromatography, also known as adsorption chromatography, separates molecules based on their polar properties. Separation is achieved by differential polar interactions (hydrogen bonding and dipole interactions) of the sample molecules with the silanol groups of the support matrix. The sample molecules, which are very polar, strongly interact with the column and tend to elute at later retention times (R_t). However, molecules that are more nonpolar tend to interact weakly with the normal phase support and elute quickly from the column at earlier retention times. The common applications for normal phase chromatography include the analysis and separation unimportant, such as in structural studies, or where structure and activity are retained despite the high concentrations of organic solvents used. This is the most common form of HPLC as most samples routinely being analyzed are water soluble. This liquid chromatography technique has become very popular for the analysis of small proteins ($\leqslant$ 50 kDa), peptides, oligonucleotides, amino acids, pharmaceuticals, pesticides, and other small organic compounds.

9 Affinity chromatography and chiral chromatography

These two techniques separate molecules according to their specific activity. Both activated and prepared supports are used for affinity chromatography and chiral chromatography. The activated supports are capable of covalently binding a molecule. Prepared supports, for example protein A supports, contain a ligand already coupled to the support, thereby making the column ready to use for the purification of molecules recognizable by the ligand. As with both an activated support and a prepared support, the selectivity and binding properties of the ligand and the molecule of interest are unique to that pair. The immobilized ligand will interact only with molecules that can selectively bind to it. Molecules that do not bind elute unretained. The retained molecules are later released in a purified state. While affinity chromatography can be used to purify biologically active molecules, such as antibodies, antigens, enzymes, and substrates, chiral chromatography can be used to purify optically active molecules, such as amino acids, drugs, pesticides, and various other small molecules.

Table 1. Typical characteristics of conventional and high-performance liquid chromatography

	Conventional (low-pressure)	High-performance (high-pressure)
Matrix particle size (μm)	100	10
Sample volume	Variable (ml–liters)	Small (μl–ml)
Flow-rate (ml cm^{-2} h^{-1})	10–30	100–300
Operating pressure (bar)	< 5	> 50
Separation time (h)	Up to 24	1–3
Resolution	Good	Can be excellent

Chapter 2 COLUMNS AND INSTRUMENTATION

1 Introduction

Efficient fractionation of molecules by liquid chromatography is made possible by the selective use of columns of appropriate support matrix and necessary instrumentation to optimize the separation process. The liquid chromatography column consists of the packing material (support matrix made up of gel or beads) and the hardware (end-fittings, frits, and tubes), with both contributing to the way in which the column performs. The column dimensions and hardware contribute to their lifetime, efficiency, and speed of analysis. The compatibility of column hardware to various solvents is summarized in *Tables 1–3*.

In addition, the instruments employed are themselves important for speed and resolution. Component accessories include pumps, automatic samplers, detectors, fraction collectors, recorders, and data acquisition systems. A guide to the different components of high-performance liquid chromatography (HPLC) systems and the necessity of the different components for separations is outlined in *Table 4* and *Table 5*, respectively. *Table 6* lists conversion factors for HPLC pump pressure.

The selection of a particular column is by far the most important step in designing and developing an analytical or purification method. It is necessary to consider the characteristics of the column, hardware and packing material, as well as the chemical and physical properties of the sample. A column can be selected based on previous experience or a literature search, or it can be selected by using a column selection guide based on known sample properties. The column matrix, as a result of the particle size, contributes to the efficiency and, through the surface chemistry, to the resolution and selectivity of the column. Different types of performance measurements are used to determine the quality and efficiency of a particular column. The most common are efficiency (plate count), resolution, and symmetry (skew). These quantities are mathematically determined from a sample chromatogram run on the column. When setting up a test procedure for monitoring column performance over its lifetime, the same procedure as that of the manufacturer should be used if possible. However, it is useful to note that a different result for the efficiency of a column will be found with each different test method, and hence it is important that each HPLC laboratory standardizes on one only. Suggested test components and appropriate test conditions for monitoring column performance and quality are outlined in *Table 7*.

To avoid repetition, a list of manufacturers and/or suppliers of liquid chromatography columns, hardware, instrumentation, and systems has not been included here; see Chapter 3 as many if not all of the manufacturers and/or suppliers listed for liquid chromatography media also market suitable columns, hardware, instrumentation, and systems.

Table 1. Solvent compatibility of polymer HPLC fittings[a]

Solvent	Polymers commonly used in HPLC[b]			
	KEL-F (CTFE)	PEEK	TEFLON (PTFE, FEP)	TEFZEL (ETFE)
Aldehydes	+	+	+	+
Aliphatic solutions	+	+	+	+
Amines	+	+	+	+
Aromatics	+	+	+	+
Bases	+	+	+	+
Chlorinated	+[c]	+	+	−
Ethers	+[c]	+	+	+
Inorganic acids	+	+[d]	+	+
Ketones	+	+	+	+
Organic acids	+	+	+	+

[a]Although every effort has been made to ensure the accuracy of this information, it is only advisory.

[b]+, recommended; −, not recommended.

[c]Not recommended for CCl_4 or THF.

[d]Not recommended for concentrated H_2SO_4 or nitric acid.

Table 2. Tubing compatibility[a]

Fluid	Tubing[b,c]			Fluid	Tubing[b,c]		
	F	S	V		F	S	V
Acetaldehyde	U	X	U	Carbon tetrachloride	X	U	T
Acetates (low mol. wt)	U	T	U	Chloroacetic acid	U	–	U
Acetic acid (less than 5%)	X	X	X	Chlorine (wet)	X	U	X
Acetic acid (more than 5%)	T	X	X	Chlorine (dry)	U	U	X
Acetic anhydride	U	U	T	Chlorobenzene	X	–	U
Acetic nitrile	–	T	–	Chloroform	X	U	T
Acetone	U	X	U	Chromic acid	X	U	X
Acetyl bromide	–	–	U	Chromium salts	–	–	X
Acetyl chloride	–	–	U	Copper salts	X	X	X
Air	X	X	X	Cresol	X	X	U
Alcohols	X	X	X	Cyclohexane	X	U	–
Aliphatic hydrocarbons	U	T	X	Diacetone alcohol	U	U	–
Aluminum chloride	X	T	X	Dimethyl formamide	U	T	–
Aluminum sulfate	X	X	X	Essential oils	–	–	X
Ammonia (gas, liquid)	U	X	X	Ethanol	X	T	T
Ammonia acetate	–	–	X	Ethers	T	U	T
Ammonium carbonate	–	–	X	Ethyl acetate	U	T	U
Ammonium chloride	X	–	X	Ethyl bromide	X	–	U
Ammonium hydroxide	X	X	X	Ethyl chlorine	X	U	U
Ammonium nitrate	–	T	X	Ethylamine	U	–	U
Ammonium phosphate	–	X	X	Ethylene glycol	X	X	X

Ammonium sulfate	X	X	X	Ethylene oxide	–	U	U
Amyl acetate	U	U	U	Fatty acids	X	T	X
Amyl alcohol	X	U	X	Ferric chloride	X	T	X
Amyl chloride	X	U	T	Ferric sulfate	X	T	X
Aniline	T	U	U	Ferrous chloride	X	T	X
Aniline hydrochloride	X	U	U	Ferrous sulfate	X	T	X
Aqua regia (75% hydrochloric				Formaldehyde	U	T	X
acid, 25% nitric acid)	T	–	U	Formic acid	U	T	X
Aromatic hydrocarbons	X	T	U	Glucose	X	X	X
Arsenic salts	X	–	X	Glycerine	X	X	X
Barium salts	X	X	X	Hydrochloric acid (all conc.[d])	X	U	X
Benzaldehyde	U	U	U	Hydrofluoric acid	X	U	X
Benzene	X	U	T	Hydrogen sulfide	U	U	X
Benzene–sulfonic acid	X	–	T	Iodine solutions	X	–	X
Benzoic acid	X	T	X	Ketones	U	U	U
Benzyl alcohol	X	–	X	Lactic acid	X	–	X
Boric acid	X	X	X	Lead acetate	U	U	X
Bromine	X	U	X	Linseed oil	X	X	X
Butane	T	U	X	Lithium hydroxide (5%)	X	U	–
Butanol	X	T	X	Magnesium chloride	X	X	X
Butyl acetate	U	–	U	Magnesium sulfate	X	–	X
Butyric acid	T	–	U	Malic acid	X	X	X
Calcium oxide (diluted)	–	X	X	Manganese salts	X	X	X
Calcium salts	X	T	X	Methane	X	U	–
Carbon bisulfide	X	–	U	Natural gas	X	X	X
Carbon dioxide	X	T	X	Nickel salts	X	X	X

Continued

Table 2. Tubing compatibility[a], *continued*

Fluid	Tubing[b,c] F	S	V
Nitric acid (diluted)	X	T	X
Nitric acid (med. conc.[d])	X	U	X
Nitric acid (conc.[d])	T	U	T
Nitrous acid	T	–	X
Oils, animal	X	T	U
Oils, mineral	X	T	U
Oils, vegetable	X	X	T
Oleic acid	T	U	T
Oxygen (gas)	–	X	X
Perchloric acid	X	U	U
Phenol	X	U	T
Phosphoric acid (ortho)	X	T	X
Potassium carbonate	X	–	X
Potassium hydroxide (med. conc. and conc.[d])	X	T	X
Pyridine	U	U	U
Silicone fluids	X	T	–
Silicone oil	X	T	T
Soap solutions	X	X	X
Sodium carbonate	X	X	X
Sodium chloride	X	X	X
Sodium hydroxide (diluted and med. conc.[d])	X	T	X
Sodium hydroxide (conc.[d])	T	T	X
Sodium hypochlorite (below 5%)	X	–	X
Sodium hypochlorite (above 5%)	X	–	T
Sodium sulfide	X	X	X
Steam (up to 40 psi)	U	U	U
Stearic acid	–	T	X
Sulfuric acid (all conc.[d])	X	U	X
Sulfurous acid	X	U	X
Tannic acid	X	T	X
Toluene	X	U	U
Trichloroacetic acid	U	–	T
Turpentine	X	U	X
Urea	X	X	X
Uric acid	–	–	X
Vinyl plastisol	–	T	–

Sodium bicarbonate	X	X	X	Water	X	X	X
Sodium bisulfate	X	–	X	Xylene	X	U	U
Sodium borate	X	X	X	Zinc chloride	X	–	X

[a]Although every effort has been made to ensure the accuracy of this information, it is only advisory.
[b]F, fluoroelastomer; S, silicone; V, vinyl.
[c]T, use only after testing; U, unsatisfactory; X, satisfactory; –, no available data.
[d]conc., concentrated; med. conc., medium concentration; all conc., all concentrations.

Table 3. Comparison of common column construction materials

Material	Biological compatibility	Chemical compatibility	Physical strength
Column wall			
Borosilicate glass	Compatible. Adsorbs protein slightly	Will not corrode	Low
Fluoropolymer	Compatible. Adsorbs protein slightly	Will not corrode	High
316 Stainless steel	Compatible. Adsorbs protein slightly	Avoid halides	Highest
End-fitting			
Fluoropolymer	Compatible	Will not corrode	High
316 Stainless steel	Compatible	Avoid halides	Highest
Titanium	Compatible	Will not corrode	High
Frits			
316 Stainless steel	Compatible. Adsorbs protein slightly	Avoid halides	High
Titanium	Compatible. Adsorbs protein slightly	Will not corrode	Moderate

Table 4. A guide to the different components of HPLC systems

Module	Component	Function	Additional information
Eluant delivery module	Solvent(s) reservoir(s)	Hold solvent in which the sample is carried through into the chromatography column; hold solvent(s) used for elution of sample components	Must hold volumes sufficient for repetition analysis; must provide for solvent de-gassing either by applying vacuum, sparging with helium, or heating; must be chemically inert with respect to the solvent
	Gradient-forming system	Prepares eluant of changing composition by mixing different solvents in a mixing chamber prior to the column	The internally generated gradient-forming system is preferred over the externally generated gradient-forming system
	Pump	Enables the chromatographer to force the eluant through the column at a constant flow-rate in order to achieve the desired separation in a short time period	Four different classes of liquid pressurizing systems exist: (i) pneumatic pumps; (ii) reciprocating piston pumps; (iii) syringe-type pumps; and (iv) pneumatic amplifier pumps. (i) and (ii) are the most popular

Sample application module		Employed to load the sample solution on to the column	Two types of sample application modules have been designed for use in HPLC: (i) micro-septum injector and (ii) septum-less microsampling valves. The latter is by far the preferred device
HPLC column module	Guard column	Protects the analytical column from contaminants originating from either the eluant or the sample	A short stainless steel column of the same internal diameter as the analytical column is used containing a pellicular support with similar chemical function to the main column. Designed to be easily replaced or re-packed with minimum delay and expense
	Analytical column	Effects the separation of the sample mixture into its various components	Either packed by the chromatographer or purchased pre-packed. Depending on type and scale of separation, different column dimensions are available
	Column ovens	Used to maintain the column temperature at a pre-set temperature (± 0.2°C), within the range 20–150°C	Aids and increases column efficiency

Continued

Table 4. A guide to the different components of HPLC systems, *continued*

Module	Component	Function	Additional information
Analyte detector module		A device through which the eluate from the HPLC column flows, and which (in most cases) generates a continuous electrical output signal. The signal is a function of the mass or of the concentration of the analyte in the mobile phase. The signal is passed directly to a chart recorder or computer system	A variety of analyte detector modules are available of which the ultraviolet (UV) photometers are most popular. See Chapter 6, *Table 1* , for a comparison of detectors
Data output module		Receives the amplified and transformed electrical response signal from the detector module	
Central control module		Controls the eluant delivery, sample application, HPLC column, analyte detector, and data output modules	

Table 5. Importance of instrumentation for HPLC separation of mixtures of small and large molecules

	Necessity of modules for the separation of molecules	
Module	Small molecules	Large molecules
Selection of appropriate HPLC column for the particular analyte mixture	Essential	Essential
Selection of appropriate guard column	Optional	Essential
HPLC column thermostat	Optional	Essential
Accurate, precise and pulse-free pump	Essential	Essential
Gradient-forming system	Optional	Essential
Suitable sample injector	Essential	Essential
Detector of suitable sensitivity	Essential	Essential
Suitable output device, e.g. strip chart recorder, chromatographic integrator, work station	Essential	Essential
Fraction collector	Optional	Optional

Table 6. HPLC pump pressure conversion

Pressure =	psi	atm	kg cm^{-2}	Torr	kPa	Bar	Inches Hg
psi =	1	0.068	0.0703	51.713	6.8948	0.06895	2.0359
atm =	14.696	1	1.0332	760	101.32	1.0133	29.921
kg cm^{-2} =	14.223	0.9678	1	735.56	98.06	0.9806	28.958
Torr =	0.0193	0.00132	0.00136	1	0.1330	0.00133	0.0394
kPa =	0.1450	0.00987	0.0102	7.52	1	0.0100	0.2962
Bar =	14.5038	0.9869	1.0197	751.88	100	1	29.5300
Inches Hg =	0.4912	0.0334	0.0345	25.400	3.376	0.03376	1

Multiply units in the left column by the conversion factors listed.
For example: to convert 250 psi to kg cm^{-2}, multiply 250 by 0.0703 = 17.57 kg cm^{-2};
to convert 50 Bar to psi, multiply 50 by 14.5038 = 725 psi.

Table 7. Test procedures for monitoring column performance

Column type	Test components[a]	Test conditions
Anion exchange	Cytidine, adenosine and uridine monophosphates	0.02 M KH_2PO_4, pH 3.5
Cation exchange	Uracil, guanine	0.02 M $NH_4H_2PO_4$, pH 3.5
Polar bonded phases (NH_2, CN, DIOL) and unbonded silicas	Naphthalene, phenyl benzoate	Chloroform:hexane (6:94)
Reversed-phase silica, C_{18} (ODS) to C_4	Benzamide, benzophenone, biphenyl	Methanol:water (70:30) or acetonitrile:water (70:30)
Reversed-phase silica, C_3 to C_1 and phenyl	Benzamide, benzophenone, biphenyl	Methanol:water (60:40)

[a]The components are listed in order of increasing retention time.

Chapter 3 CHROMATOGRAPHY MEDIA

1 Introduction

This chapter lists the different types of media used for the major types of liquid chromatography. The choice of media depends on the type of separation that is required and whether the separation is to be carried out using a high-pressure liquid chromatography (HPLC) system.

The following tables are not intended to be exhaustive and there may well be column support materials, and manufacturers and suppliers other than those listed. Where media have been given as a series the reader should consult the manufacturers for the individual types of media and their properties in that series.

The manufacturers and suppliers have been given in abbreviated form; consult Chapter 8 for the complete names and addresses.

2 Media for size exclusion chromatography

Sephadex was one of the first commercially available media for size exclusion chromatography. Sepharose and subsequently Sephacryl, which has a faster flow rate, have been used for fractionating larger molecules. The choice of media is dependent on the size of molecules to be separated (*Figure 1*). None of the media listed in *Tables 1* and *2* is suitable for HPLC. For HPLC systems, the media used are based on either silica particles or polymeric beads. *Tables 3* and *4* list some of the media that are used for size exclusion chromatography in HPLC systems.

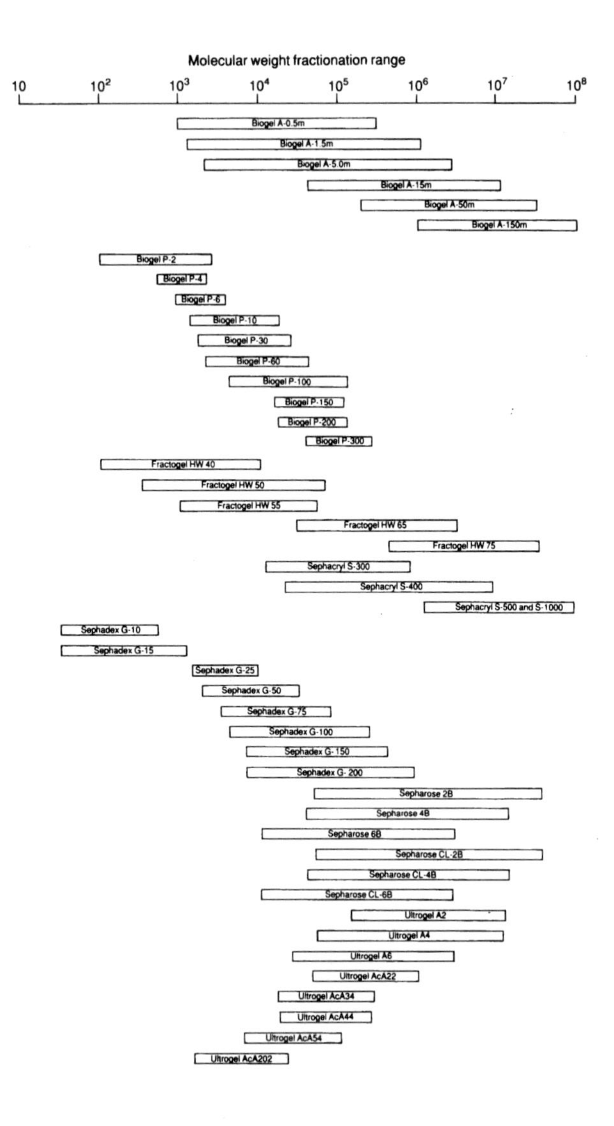

Figure 1. Fractionation ranges of commercially available gel filtration matrices. Reproduced from Patel (1993) Chromatographic frationation media, in *Biochemistry Labfax* (Chambers and Rickwood, eds), p. 51, BIOS Scientific Publishers.

3 Media for hydrophobic interaction chromatography

Hydrophobic interaction chromatography columns are packed with either silica or polymer-based macroporous particles or beads (*Table 5*). The silica-based media are stable in the pH range 4–8, while the polymer-based media are stable in the pH range 2–12.

4 Media for hydroxylapatite chromatography

High-performance hydroxylapatite columns are packed with a crystalline structure of calcium phosphate ($[Ca_5(PO_4)_3OH]_2$) (*Table 6*). The media are stable in the pH range 5.5–10.5 and at elevated temperatures up to 85°C.

5 Media for ion-exchange chromatography

Most of the original media for ion-exchange chromatography were based on polysaccharide supports (e.g. cellulose, *Table 7*). However, due to their irregular shape, many cellulose ion exchangers had low capacities and poor flow properties, and were replaced by media with improved flow properties and high capacities for macromolecules. Modern high-performance media are usually derived from silica, hydroxylated polyethers, polystyrene and cross-linked, nonporous polyacrylates. These are chemically bonded with acidic groups such as sulfonic acid or carboxylic acid for the separation of cations, or basic groups such as amine or quaternary amine for the separation of anions. *Tables 8–12* list the media available.

6 Media for ion-moderated partition chromatography

Ion-moderated partition columns are packed with a polystyrene-based support matrix (*Table 13*). Functional groups (or ligands) attached to the matrix give each column a specific selectivity.

7 Media for normal phase chromatography

Normal phase chromatography columns are packed with

porous silica particles (*Table 14*). The functional group of the column is the silanol group of the silica. However, the addition of hydroxyl groups, in the form of polyols, is not uncommon. The media are stable in the pH range 2–7.

8 Media for reversed-phase chromatography

The media for this type of separation are essentially based on silica chemically bonded with an alkylsilyl compound to give a nonpolar, hydrophobic surface. These are presented in *Table 15*. A smaller range of non-silica- and polystyrene-based media are available as shown in *Table 16*. The media are stable in the pH range 2–7 and at elevated temperatures.

9 Media for affinity chromatography

Originally supports were based on activated Sepharose but now a wide range of supports is available including media able to withstand the pressure of HPLC systems. *Table 17* summarizes the properties of some of the media that are commercially available.

10 Media for chiral chromatography

Chiral chromatography commonly uses porous silica matrices with various optically active compounds attached. A selection of the many commercially available chiral supports is presented in *Table 18*; some of these are cellulose derivatives coated on to silica gel.

Table 1. Carbohydrate-based column support materials for separation by size exclusion

Name and type of matrix	Fractionation range ($\times 10^3$)		Dry bead diameter (μm)	Approx. bed volume (ml g^{-1})	Comments
	Peptides, globular proteins	Dextrans			
Sephadex (beaded gel of dextran cross-linked with epichlorohydrin)					
G-10	< 0.7	< 0.7	40–120	2–3	pH range, 2–12; temperature stability, 120°C; stability to solvents, avoid strong acid and oxidizing agents; subject to microbial degradation
G-15	< 1.5	< 1.5	40–120	2.5–3.5	
G-25 coarse	1–5	0.1–5	100–300	4–6	
G-25 medium	1–5	0.1–5	50–150	4–6	
G-25 fine	1–5	0.1–5	20–80	4–6	
G-25 superfine	1–5	0.1–5	20–50	4–6	
G-50 coarse	1.5–30	0.5–10	100–300	9–11	
G-50 medium	1.5–30	0.5–10	50–150	9–11	
G-50 fine	1.5–30	0.5–10	20–80	9–11	
G-50 superfine	1.5–30	0.5–10	20–50	9–11	
G-75	3–80	1–50	40–120	12–15	
G-75 superfine	3–70	1–50	20–50	12–15	
G-100	4–150	1–100	40–120	15–20	
G-100 superfine	4–100	1–100	20–50	15–20	
G-150	5–300	1–150	40–120	20–30	
G-150 superfine	5–150	1–150	20–50	18–22	
G-200	5–600	1–200	40–120	30–40	

Continued

Table 1. Carbohydrate-based column support materials for separation by size exclusion, *continued*

Name and type of matrix	Fractionation range ($\times 10^3$)		Dry bead diameter (μm)	Approx. bed volume (ml g^{-1})	Comments
	Peptides, globular proteins	Dextrans			
G-200 superfine	5–250	1–200	20–50	20–25	
PDX					
(cross-linked dextran beads)					
GF-25	1–5	–	50–150	4–6	
GF-25	1–5	–	100–300	4–6	
GF-50	1.5–30	–	50–150	9–11	
GF-50	1.5–30	–	100–300	9–11	
Sephacryl					
(cross-linked copolymer of allyl dextran and *N,N′*-methylenebis(acrylamide))					
S-100 high resolution	1–100	–	25–75	Pre-swollen	pH range, 2–11; temperature stability, 120°C; stable in solvents
S-200 superfine	5–250	1–80	40–105	Pre-swollen	
S-200 high resolution	5–250	1–80	25–75	Pre-swollen	
S-300 superfine	10–1500	1–400	40–105	Pre-swollen	
S-300 high resolution	10–1500	1–400	25–75	Pre-swollen	
S-400 superfine	20–8000	10–2000	40–105	Pre-swollen	
S-400 high resolution	20–8000	10–2000	25–75	Pre-swollen	
S-500 superfine	$40–20 \times 10^3$	–	40–105	Pre-swollen	
S-500 high resolution	$40–20 \times 10^3$	–	25–75	Pre-swollen	
S-1000 superfine	$5000–10^4$	–	40–105	Pre-swollen	

Sepharose and Sepharose CL (beaded agarose)					
6B	10–4000	10–1000	45–165	Pre-swollen	Sepharose: pH range, 4–9; stable at 40°C; avoid urea, organic solvents and chaotropic salts
6B cross-linked	10–4000	10–1000	45–165	Pre-swollen	Sepharose CL: pH range, 3–14; stable at 120°C; avoid strong oxidizing agents
4B	$60–20 \times 10^3$	30–5000	60–140	Pre-swollen	
4B cross-linked	$60–20 \times 10^3$	30–5000	60–140	Pre-swollen	
2B	$70–40 \times 10^3$	$100–20 \times 10^3$	60–200	Pre-swollen	
2B cross-linked	$70–40 \times 10^3$	$100–20 \times 10^3$	60–200	Pre-swollen	
Superose (highly cross-linked agarose matrix)					
6 Prep. grade	5–5000	–	20–40	Pre-swollen	
12 Prep. grade	1–300	–	20–40	Pre-swollen	
Trisacryl Plus (beaded poly(*N*-tris[hydroxymethyl]methyl methacrylamide))					
GF2-M	1–15	–	40–80	Pre-swollen	
GF4-M	5–25	–	40–80	Pre-swollen	
GF10-M	5–65	–	40–80	Pre-swollen	
GF20-M	10–130	–	40–80	Pre-swollen	

Supplier: Sigma (SCC).

Table 2. Controlled pore glass for permeation chromatography

Nominal pore size (Å)	Unmodified, mesh[a]	Glyceryl-controlled pore glass, mesh
75	120–200	120–200
75	200–400	200–400
120	120–200	120–200
120	200–400	200–400
170	120–200	120–200
170	200–400	200–400
240	120–200	120–200
240	200–400	200–400
350	120–200	120–200
350	200–400	200–400
500	120–200	120–200
500	200–400	200–400
700	120–200	120–200
700	200–400	200–400
1000	120–200	120–200
1000	200–400	200–400
1400	120–200	120–200
1400	200–400	200–400

2000	120–200	120–200
2000	200–400	200–400
3000	80–120	120–200
3000	120–200	200–400

[a]Most beads are also available in 20–80 and 80–120 mesh.
Manufacturer: Electro-Nucleonics (ENI).

Table 3. Silica-based column support materials available for size exclusion HPLC of proteins and peptides

Matrix	Pore size (nm)	Mol. wt $\times 10^3$ exclusion limit	Particle size (μm)	Column sizes: Length (cm)	Column sizes: i.d. (mm)	Availability bulk material	Manufacturer
Aquapore OH-100	10	90	10	25	4.6	No	PEL
Aquapore OH-300	30	800	10	25	4.6	No	PEL
Aquapore OH-500	50	5000	10	25	4.6	No	PEL
Aquapore OH-1000	100	20×10^3	10	25	4.6	No	PEL
Bio-Sil SEC 125	12.5	5–100	5, 10, 13	8, 30, 60	7.5, 7.8, 21.5	–	BRL
Bio-Sil SEC 250	25	10–300	5, 10, 13	8, 30, 60	7.5, 7.8, 21.5	–	BRL
Bio-Sil SEC 400	40	20–1000	5, 13, 17	8, 30, 60	7.5, 7.8, 21.5	–	BRL
LiChrospher DIOL	10	70	5, 7	25	4.0	Yes	MBH
LiChrospher DIOL	50	800	10	25	4.0	No	MBH
LiChrospher DIOL	100	3000	10	25	4.0	No	MBH

Continued

Table 3. Silica-based column support materials available for size exclusion HPLC of proteins and peptides, *continued*

Matrix	Pore size (nm)	Mol. wt $\times 10^3$ exclusion limit	Particle size (μm)	Column sizes		Availability bulk material	Manufacturer
				Length (cm)	i.d. (mm)		
Polyol Si300	30	600	3, 5, 10	25, 50	4.6, 7.1, 9.5	Yes	SFG
Polyol Si500	50	1000	10	25, 50	4.6, 7.1, 9.5	Yes	SFG
SynChropak 60	6	30	5	25, 30	4.6, 7.8	Yes	SCI
SynChropak 100	10	300	5	25, 30	4.6, 7.8	Yes	SCI
SynChropak 300	30	1500	5	25, 30	4.6, 7.8	Yes	SCI
SynChropak 500	50	10×10^3	7	25, 30	4.6, 7.8	Yes	SCI
SynChropak 1000	100	50×10^3	7	25, 30	4.6, 7.8	Yes	SCI
SynChropak 4000	400	–	10	25, 30	4.6, 7.8	Yes	SCI
TSK 2000 SW	12.5	70	10	30, 50	7.5, 21.5	No	THS
TSK 3000 SW	25	300	10	30, 60	7.5, 21.5	No	THS
TSK 4000 SW	50	1000	13	30	7.5	No	THS
Zorbax Bio GF-250	25	400	4	25	9.4	No	DPL

i.d., internal diameter.

Table 4. Polymer-based column support materials available for size exclusion HPLC of proteins and peptides

Matrix	Pore size (nm)	Mol. wt $\times 10^3$ exclusion limit	Particle size (μm)	Column sizes		Availability bulk material	Manufacturer
				Length (cm)	i.d. (mm)		
Bio-Gel SEC 10	< 10	< 2	10	30	7.5	–	BRL
Bio-Gel SEC 20XL	< 20	< 8	6	30	7.5	–	BRL
Bio-Gel SEC 30XL	20	0.5–800	6	30	7.5	–	BRL
Bio-Gel SEC 40XL	50	10–1500	10	30	7.5	–	BRL
Bio-Gel SEC 50XL	100	$< 1 \times 10^4$	10	30	7.5	–	BRL
Bio-Gel SEC 60XL	> 100	$< 20 \times 10^4$	13	30	7.5	–	BRL
MCI Gel CQS10	10	400	9–11	30	7.5	Yes	MCI
MCI Gel CQS30	30	1000	9–11	30	7.5	Yes	MCI
MCI Gel CQP06	6	1	9–11	30	7.5	Yes	MCI
MCI Gel CQP10	10	400	9–11	30	7.5	Yes	MCI
MCI Gel CQP30	30	1000	7–10	30	7.5	Yes	MCI
Micropak TSK Gel	4–1000	1–400	10	–	–	–	VUL
H Series	$1 \times 10^4 - 1 \times 10^6$	$4000–40 \times 10^4$	17	–	–	–	VUL
PLgel Series	$5–1 \times 10^5$	$0.5–40 \times 10^3$	5, 10	30	7.5	Yes	PLL
PLgel MIXED-A		$1–40 \times 10^3$	20	30	7.5	–	PLL
PLgel MIXED-B		$0.5–10 \times 10^3$	10	30	7.5	–	PLL
PLgel MIXED-C		0.2–3000	10	30	7.5	–	PLL
PLgel MIXED-D		0.2–400	5	30	7.5	–	PLL
PLgel MIXED-E		Up to 30	3	30	7.5	–	PLL

Continued

Table 4. Polymer-based column support materials available for size exclusion HPLC of proteins and peptides, *continued*

Matrix	Pore size (nm)	Mol. wt $\times 10^3$ exclusion limit	Particle size (μm)	Column sizes		Availability bulk material	Manufacturer
				Length (cm)	i.d. (mm)		
TSK 2000 PW	5	4	10	30, 50	7.5, 21.5	No	THS
TSK 3000 PW	20	50	13	30, 60	7.5, 21.5	No	THS
TSK 4000 PW	50	300	13	30, 60	7.5, 21.5	No	THS
TSK 5000 PW	100	1000	17	30, 50	7.5, 21.5	No	THS
TSK 6000 PW	> 100	8000	25	30, 50	7.5, 21.5	No	THS

i.d., internal diameter.

Table 5. Selection of column supports for hydrophobic interaction chromatography

Name	Functional group	Particle size (μm)	Column dimensions length (cm) × i.d. (mm)	Manufacturer
Silica-based columns				
Bakerbond HI-Propyl	$CH_2CH_2CH_3$	5	25 × 4.6	JTB
Poly Propyl A	$CH_2CH_2CH_3$	5	10, 20 × 4.6	PLC
Poly Ethyl A	CH_2CH_3	5	10, 20 × 4.6	PLC
Poly Methyl A	CH_3	5	10, 20 × 4.6	PLC
SynChropak Propyl	$(CH_2)_2CH_3$	6.5	25, 30 × 4.6, 10	SCI
SynChropak Hydroxy	$(CH_2)_2CH_2OH$	6.5	25, 30 × 4.6, 10	SCI
SynChropak Methyl	CH_3	6.5	25, 30 × 4.6, 10	SCI
SynChropak Pentyl	$(CH_2)_4CH_3$	6.5	25, 30 × 4.6, 10	SCI
Polymer-based columns				
Bio-Gel MP7 HIC	Methyl	7	5 × 7.8	BRL
Bio-Gel Phenyl-5-PW	Phenyl	10, 15	7.5 × 7.5, 15 × 21.5	BRL
POROS BU	Butyl	10, 20, 50		PSB
POROS ET	Ether	10, 20, 50		PSB
POROS HP	Phenyl	10, 20, 50		PSB
POROS PE	Phenyl ether	10, 20, 50		PSB
POROS PH	Phenyl	10, 20, 50		PSB
Progel-TSK Butyl-NPR	Butyl	2.5	3.5 × 4.6	SUK
Progel-TSK Ether-5PW	Ether	10	7.5 × 7.5	SUK
Progel-TSK Phenyl-5PW	Phenyl	10	7.5 × 7.5	SUK
TSK Gel Ether-5PW	Oligo-ethyleneglycol	10	7.5 × 7.5, 15 × 21.5	THS
TSK Gel Phenyl-5PW	Phenyl	10	7.5 × 7.5, 15 × 21.5	THS

i.d., internal diameter.

Table 6. Commercially available hydroxylapatite columns for HPLC

Name	Particle size (μm)	Column dimensions length (cm) × i.d. (mm)	Manufacturer
Bio-Gel HPHT	–	10 × 7.8	BRL
HCA-Column (A-7610)	–	7.5 × 4, 10 × 7.6	MTC
Pentax	2, 10, 20	1, 5, 10 × 4.0, 7.5	PXG
Progel-TSK HA-1000	5	7.5 × 75	SUK
TSK Gel HA-1000	5	7.5 × 7.5	THS
HA-Ultrogel	60–180	–	LSL

i.d., internal diameter.

Table 7. Ion-exchange cellulose media: physical and chemical properties of Whatman cellulose media

Physical form	Functional group	Normal pH range	Small ion capacity (meq dg^{-1})	Protein capacity (mg dg^{-1})	Bed volume (mg ml^{-1})	Amount of exchanger required per liter bed volume (kg)	Packing density (dg ml^{-1})
Anion exchange media							
Pre-swollen microgranular							
DE-51	Diethylaminoethyl	2–9	0.20–0.25	175	30	1.20	0.17
DE-52	Diethylaminoethyl	2–9.5	0.88–1.08	700	130	0.90	0.19

DE-53	Diethylaminoethyl	2–12	1.8–2.2	750	150	1.05	0.20
CA-52	Quaternary ammonium	2–12	1.0–1.2	750	150	1.20	0.20
Dry microgranular							
DE-32	Diethylaminoethyl	2–9.5	0.88–1.08	700	140	0.24	0.20
Dry fibrous							
DE-23	Diethylaminoethyl	2–9.5	0.88–1.08	425	60	0.19	0.15
Cation exchange media							
Pre-swollen microgranular							
CM-52	Carboxymethyl	3–10	0.90–1.15	1180	210	1.05	0.18
SE-52	Sulfoxyethyl	2–12	0.9–1.1	1300	195	1.05	0.15
SE-53	Sulfoxyethyl	2–12	2.1–2.6	1300	210	1.15	0.16
Dry microgranular							
CM-32	Carboxymethyl	3–10	0.90–1.15	1180	200	0.21	0.17
Dry fibrous							
CM-23	Carboxymethyl	3–10	0.55–0.70	675	85	0.16	0.13
P-11	Orthophosphate	3–10	3.2–5.3	–	–	0.22	0.17
Hydrogen bonding medium							
Pre-swollen microgranular							
HB-1	Hydroxyl	2–12	0	375	–	0.70	0.15
Cell debris remover							
Pre-swollen fibrous							
CDR	Diethylaminoethyl	2–9.5	0.25–0.35	–	–	1.15	0.19

dg, dry gram.

Table 8. Analytical grade resins for ion-exchange chromatography

Resin	Type	Matrix	Cross-linking (%)	Ionic form	Mol. wt $\times 10^3$ exclusion limit	Solvent compatibility
AG 1	Strong anion	Styrene divinylbenzene	2	Acetate, chloride	2.7	Very good
			4	Chloride	1.4	
			8	Acetate, chloride, formate, hydroxide	1.0	
AG 2	Strong anion	Styrene divinylbenzene	8	Chloride	1.0	Very good
AG 3-X4	Weak anion	Polyamine	4	Free base	1.4	Good
AG 4-X4	Weak anion	Acrylic	4	Free base	1.4	Good
AG 11 A8	Ion retardation	Acrylic acid within styrene divinylbenzene	8	Self adsorbed	1.0	Good
AG 50W	Strong cation	Styrene divinylbenzene	2	Hydrogen	2.7	Very good
			4	Hydrogen	1.4	
			8	Hydrogen	1.0	
			12	Hydrogen	0.4	
AG 501-X8	Mixed bed	Styrene divinylbenzene	8	$H^+ + OH^-$	1.0	Very good
Bio-Rex 5	Weak anion	Styrene divinylbenzene	4	Chloride	1.4	Good
Bio-Rex 70	Weak cation	Acrylic	Macroreticular	Sodium	> 75	Good
Bio-Rex MSZ 501	Mixed bed	Styrene divinylbenzene	8, 10	$H^+ + OH^-$	1.0	Very good
Chelex 100	Weak cation	Styrene divinylbenzene	1	Iron, sodium	3.5	Good

Table 9. Ion exchange: anion exchangers on polystyrene supports

Cross-linkage (%)	Gel or macroret[a]	Mesh size	Ionic form	Moisture content (%)	Max. op. temp. (°C)	Total exchange capacity (meq ml^{-1})	Total exchange capacity (meq g^{-1})	pH range
Amberlite weakly basic anion exchangers								
–	M	16–50	Free base	57	100	1.2	4.7	0–7
–	M	16–50	Free base	60	100	1.2	4.7	0–7
–	G	16–50	Free base	60	60	1.6	5.6	0–7
Amberlite strongly basic anion exchangers								
–	M	16–50	Cl^-	59	60 (OH); 80 (Cl)	1.0	4.2	0–14
–	M	20–50	Cl^-	57	60 (OH); 80 (Cl)	0.6	2.3	0–14
8	G	16–50	Cl^-	46	60 (OH); 80 (Cl)	1.4	3.8	0–14
8	G	16–50	OH^-	46	60 (OH); 80 (Cl)	1.15	4.0	0–14
8	G	16–45	Cl^-	46	60 (OH); 80 (Cl)	1.4	3.8	0–14
6	G	16–50	Cl^-	53	60 (OH); 80 (Cl)	1.25	4.1	0–14
–	G	20–50	Cl^-	42	40 (OH); 80 (Cl)	1.4	3.4	0–14
–	G	16–45	Cl^-	53	60 (OH); 80 (Cl)	1.25	4.1	0–14
–	G	16–50	Cl^-	58	60 (OH); 80 (Cl)	1.2	4.4	0–14
–	G	100–200	Cl^-	50–58	60 (OH); 80 (Cl)	1.3	1.8	0–14
Dowex 1 strongly basic anion exchangers								
2	G	50–100	Cl^-	65–75	150	0.7	3.5	0–14
2	G	100–200	Cl^-	70–80	150	0.6	3.5	0–14
2	G	200–400	Cl^-	70–80	150	0.6	3.5	0–14
4	G	20–50	Cl^-	$\geqslant$ 50	150	1.0	3.5	0–14
4	G	50–100	Cl^-	$\geqslant$ 50	150	1.0	3.5	0–14

Continued

Table 9. Ion exchange: anion exchangers on polystyrene supports, *continued*

Cross-linkage (%)	Gel or macroret[a]	Mesh size	Ionic form	Moisture content (%)	Max. op. temp. (°C)	Total exchange capacity		pH range
						(meq ml^{-1})	(meq g^{-1})	
4	G	100–200	Cl^-	55–63	150	1.0	3.5	0–14
4	G	200–400	Cl^-	55–63	150	1.0	3.5	0–14
8	G	20–50	Cl^-	43–48	150	1.2	3.5	0–14
–	G	20–50	OH^-	$\leqslant$ 60	50	1.1	3.9	0–14
–	G	16–40	OH^-	60–70	50	1.0	4.4	0–14
8	G	50–100	Cl^-	43–48	150	1.2	3.5	0–14
8	G	100–200	Cl^-	39–45	150	1.2	3.5	0–14
8	G	200–400	Cl^-	39–45	150	1.2	3.5	0–14
8	G	50–100	Cl^-	$\geqslant$ 38	150	1.2	–	0–14
8	G	100–200	Cl^-	34–40	150	1.2	–	0–14
8	G	200–400	Cl^-	34–40	150	1.2	–	0–14

Max. op. temp., maximum operating temperature.

[a]Gel is in the form of beads where ion exchange occurs on the surface of the beads while macroret is a porous matrix where ion exchange occurs in the pores.

Supplier: Sigma (SCC).

Table 10. Ion exchange: cation exchangers on polystyrene supports

Cross-linkage (%)	Gel or macroret[a]	Mesh size	Ionic form	Moisture content (%)	Max. op. temp. (°C)	Total exchange capacity (meq ml^{-1})	Total exchange capacity (meq g^{-1})	pH range
Amberlite weakly acidic cation exchangers								
4	M	16–50	Na^+	67	120	2.5	8.1	5–14
4	M	16–50	H^+	48	120	3.5	10.0	5–14
4	M	100–200	H^+	5	120	3.5	10.0	5–14
4	M	100–500	H^+	10	120	3.5	10.0	5–14
4	M	100–500	K^+	10	120	2.5	10.0	5–14
Amberlite strongly acidic cation exchangers								
20	M	16–50	Na^+	48	150	1.7	4.2	0–14
4.5	G	16–50	H^+	65	120	1.3	4.9	0–14
8	G	16–50	Na^+	45	120	1.9	4.4	0–14
10	G	16–50	Na^+	40	120	2.1	–	0–14
–	G	16–50	Na^+	46	120	1.9	4.4	0–14
8	G	100–200	Na^+	8	120	1.9	4.4	0–14
8	G	100–500	Na^+	10	120	1.9	4.3	0–14
Dowex 50W strongly acidic cation exchangers								
1	G	50–100	H^+	51–54	150	1.8	4.8	0–14
2	G	50–100	H^+	74–82	150	0.6	4.8	0–14
2	G	100–200	H^+	74–82	150	0.6	4.8	0–14
2	G	200–400	H^+	74–82	150	0.6	4.8	0–14
4	G	50–100	H^+	64–72	150	1.1	4.8	0–14

Continued

Table 10. Ion exchange: cation exchangers on polystyrene supports, *continued*

Cross-linkage (%)	Gel or macroret[a]	Mesh size	Ionic form	Moisture content (%)	Max. op. temp. (°C)	Total exchange capacity (meq ml^{-1})	Total exchange capacity (meq g^{-1})	pH range
4	G	100–200	H^+	64–72	150	1.1	4.8	0–14
4	G	200–400	H^+	64–72	150	1.1	4.8	0–14
8	G	20–50	H^+	51–54	150	1.8	4.8	0–14
8	G	50–100	H^+	50–56	150	1.7	4.8	0–14
8	G	100–200	H^+	50–58	150	1.7	4.8	0–14
8	G	200–400	H^+	50–58	150	1.7	4.8	0–14
12	G	50–100	H^+	51–54	150	1.8	4.8	0–14

Max. op. temp., maximum operating temperature.

[a]Gel is in the form of beads where ion exchange occurs on the surface of the beads while macroret is a porous matrix where ion exchange occurs in the pores.

Supplier: Sigma (SCC).

Table 11. Some commercially available column support materials for ion-exchange HPLC of proteins

Name	Surface	Functional group	Pore size (nm)	Particle size (μm)	Column dimensions length (cm) × i.d. (mm)	Manufacturer/ supplier
Silica-based columns						
Aquapore AX-300	Weak anion	Diethylaminoethyl	30	7	3, 10, 22 × 2.1, 4.6	PRC, PEL
Aquapore AX-1000	Weak anion	Diethylaminoethyl	10	10	10, 22 × 4.6, 25 × 7.0	PEL
Aquapore CX-300	Weak cation	Carboxymethyl	30	7	3, 10, 22 × 2.1, 4.6	PRC, PEL
Aquapore CX-1000	Weak cation	Carboxymethyl	10	10	10, 22 × 4.6, 25 × 7.0	PEL
Bakerbond CBX	Weak anion	Carboxyethyl	30	5	5 × 4.6	JTB
Bakerbond PEI	Weak anion	Polyethyleneimine, CH_2CH_2NH	30	5	5 × 4.6	JTB
Daltosil			30, 10	3, 5, 10		SFG
Dynamax ~ 300 Å AX			30	5, 12		RIC
Exsil 100 SAX	Strong anion		10	3, 5, 10, 15	5, 10, 15, 25 × 4.6	HCL
Exsil 100 SCX	Strong cation		10	3, 5, 10, 15	5, 10, 15, 25 × 4.6	HCL
LiChrosorb NH_2		Primary amino (bonded)	6	5		MHB
Nucleogen 60	Weak anion	Tertiary amino (bonded)	6	7		MNG
Nucleogen 500	Weak anion	Tertiary amino (bonded)	50	10		MNG
Nucleogen 4000	Weak anion	Tertiary amino (bonded)	400	10		MNG
Nucleosil AN/CAT			10	5, 10		MNG

Continued

Table 11. Some commercially available column support materials for ion-exchange HPLC of proteins, *continued*

Name	Surface	Functional group	Pore size (nm)	Particle size (μm)	Column dimensions length (cm) × i.d. (mm)	Manufacturer/ supplier
Partisil SAX		Quaternary amino (bonded)	10	10		WIL
PEI widepore		Primary and secondary amino (pellicular)	33	5		JTB
Supelcosil			30	5		SUK
SynChropak AX300	Weak anion	Polyamine	30	6.5	10, 25 × 4.6, 10, 22.5	SCI
SynChropak AX1000	Weak anion		30	10	10, 25 × 4.6, 10, 22.5	SCI
SynChropak CM300	Weak anion	Carboxymethyl	30	6.5	10, 25 × 4.6, 10	SCI
SynChropak Q300	Strong anion	Quaternary amine	30	6.5	10, 25 × 4.6, 10, 22.5	SCI
SynChropak Q1000	Strong anion	Quaternary amine	30	10	10, 25 × 4.6, 10, 22.5	SCI
SynChropak S300	Strong cation	Sulfonic acid	30	6.5	10, 25 × 4.6, 10	SCI
TSK CM-2/SW	Weak cation	Carboxyl	13	5	30 × 4.6	THS
TSK CM-3/SW	Weak cation	Carboxyl	25	10	15 × 6.0, 21.5	THS
TSK CM-5/SW	Weak cation	Carboxyl	50	10	15 × 4.6, 21.5	THS
TSK DEAE-2/SW	Weak anion	$-N^{+}HEt_2$	13	5	30 × 4.6	THS
TSK DEAE-3/SW	Weak anion	$-N^{+}HEt_2$	25	10	15 × 6.0, 21.5	THS
Waters Accell	Cation	Carboxymethyl	50	37–55	Bulk material only	PSL
Zorbax 300 SCX	Strong cation	Sulfonic acid	5, 7	30		DPL
Zorbax SAX	Strong anion	Quaternary ammonium	5, 7	7		DPL

Polymer-based columns						
Bio-Gel MA7C		Carboxyl	Nonporous	7	3 × 4.6, 5 × 7.8, 10 × 19	BRL
Bio-Gel MA7S		Sulfopropyl	Nonporous	7	5 × 7.8, 10 × 19	BRL
Bio-Gel MA7P		Polyethyleneimine	Nonporous	7	3 × 4.6, 5 × 7.8, 10 × 19	BRL
Bio-Gel MA7Q		Quaternary amine	Nonporous	7	5 × 7.8, 10 × 19	BRL
Mono Q	Anion	Quaternary amine	70	10	50 × 5	PMB
Mono S	Cation	Sulfonate	–	10	50 × 5	PMB
PL SAX	Strong anion	Quaternary amine	100	8, 10	5, 15 × 4.6, 5, 15 × 7.5, 25	PLL
POROS CM	Weak cation	Carboxymethyl	600–800	10, 20, 50		PSB
POROS HQ	Strong anion	Quaternized polyethyleneimine	600–800	10, 20, 50		PSB
POROS PI	Weak anion	Polyethyleneimine	600–800	10, 20, 50		PSB
POROS Q	Strong anion	Quaternized polyethyleneimine	600–800	10, 20, 50		PSB
POROS S	Strong cation	Sulfoethyl	600–800	10, 20, 50		PSB
POROS SP	Strong cation	Sulfopropyl	600–800	10, 20, 50		PSB
TSK DEAE 5 PW	Weak anion	$-N^{+}HEt_2$	100	10	7.5 × 7.5, 15 × 21.5	THS
TSK Gel DEAE-NPR	Weak anion	$-N^{+}HEt_2$	100	2.5	3.5 × 4.6	THS
TSK Gel SP-NPR	Weak cation	$-(CH_2)_3SO_3^-$	100	2.5	3.5 × 4.6	THS
TSK SP 5 PW	Strong cation	$-SO_3^-$	100	10	7.5 × 7.5, 15 × 21.5	THS

i.d., internal diameter.

Table 12. Some commercially available column support materials for cation-exchange HPLC of oligosaccharides

Name	Ionic form	Cross-linking (%)	Particle size (μm)	Manufacturer/supplier
Animex HPX-87N	Na^{+}	8	9	BRL
Animex HPX-87K	K^{+}	8	9	BRL
Animex HPX-42C	Ca^{2+}	4	25	BRL
Animex HPX-42A	Ag^{+}	4	25	BRL
Animex HPX-65A	Ag^{+}	6	11	BRL
Bio-Rad Fast Carbohydrate	$Pb^{2+/4+}$	8	9	BRL
Interaction CHO-411	Na^{+}	–	20	FSL
Interaction CHO-611	Na^{+}	–	12	FSL
Interaction CHO-682	$Pb^{2+/4+}$	–	8	FSL
MCI Gel CK02A	–	2	20–24	MCI
MCI Gel CK04S	–	4	11–14	MCI
MCI Gel CK04C	–	4	17–20	MCI
MCI Gel CK06S	Na^{+}	6	11–14	MCI
Polypore CA	Ca^{2+}	–	10	PEL
Spherogel Carbohydrate	Ca^{2+}	–	10	BKI

Table 13. Various ion-moderated partition chromatography columns

Name	Ionic form	Cross-linking (%)	Particle size (μm)	Column dimension length (cm) × i.d. (mm)	pH range
Animex HPX-42A	Ag^{+}	4	25	30 × 7.8	6–8
Animex HPX-42C	Ca^{2+}	4	25	30 × 7.8	5–9
Animex HPX-72S	SO_4^{2-}	7	11	30 × 7.8	4–9
Animex HPX-87C	Ca^{2+}	8	9	25, 30 × 4.0, 7.8	5–9
Animex HPX-87H	H^{+}	8	9	30 × 7.8	1–3
Animex HPX-87K	K^{+}	8	9	30 × 7.8	5–9
Animex HPX-87N	Na^{+}	8	9	30 × 7.8	5–9
Animex HPX-87P	$Pb^{2+/4+}$	8	9	30 × 7.8	5–9
Fast Acid	H^{+}	8	9	10 × 7.8	1–3
Fast Carbohydrate	$Pb^{2+/4+}$	8	9	10 × 7.8	5–9

Manufacturer: Bio-Rad (BRL).

Table 14. List of normal phase column support materials

Name	Particle diameter (μm)	Mean pore diameter (nm)	Supplier
Bio-Sil polyol	10	9	BRL
Chromegabond amine	3, 5, 10	6, 8, 10	HCL
Chromegabond diol	3, 5, 10	6, 10	HCL
Chromegabond nitrile (cyano)	3, 5, 10	6, 8, 10	HCL
Chromegabond amino/cyano	5, 10	6	HCL
Hypersil	3	12	SSP
Hypersil APS	3	12	SSP
Hypersil CPS	3	12	SSP
Nova-Pak silica	4	6	PSL
Nucleosil CN	5, 7	10, 12	MNG
Nucleosil NH_2	5, 7	10, 12	MNG
Nucleosil OH	7	10	MNG
Partisil silica	5, 10	8.5	WIL
PVA-Sil	5	12	HCL
Resolve Silica	5	9	PSL
Spheri-5 amino	5	8	PRC
Spheri-5 cyano	5	8	PRC
Spheri-5 silica	5	8	PRC
Spherisorb S3W	3	8	PSL
Spherisorb S5W	5	8	PSL
Spherisorb S10W	10	8	PSL
Ultrasphere-CN	5	–	BKI
Ultrasphere-Si	5	–	BKI
Zorbax CN	3, 5, 7	7	HCL
Zorbax NH_2	3, 5, 7	7	HCL

Table 15. List of silica-based reversed-phase column support materials

Name	Particle diameter (μm)	Mean pore diameter (nm)	Manufacturers/ suppliers
Bonded phase: C_1			
Astec 300 A C1[a]	5	30	BTL
Spherisorb S5(10)X C1[a]	5 (10)	30	PSL
SynChropak RP-1[a]	6.5	100	SCI
TSK Gel TMS-250	10	25	THS
Ultropak TSK TMS-250	10	25	PMB
Bonded phase: C_3			
Bakerbond widepore HI propyl[a]	5	30	JTB
Ultrapore RPSC	5	30	ACL, BKI
Bonded phase: C_4			
Apex WP Butyl[a]	7	30	JCL
Aquapore BU-300	7	30	ACL, PRC
Aquapore RP-300	7	30	PRC, PEL
Astec 300A C4[a]	5	30	BTL
Bakerbond widepore Butyl[a]	5	30	JTB
Hipore RP-304	5, 10	30	BRL
Hypersil WP Butyl[a]	5	30	SSP
Nucleosil 300 C4	5, 7, 10	30	MNG
Supelco LC-304	5	30	SUK
SynChropak RP-4[a]	6.5	30, 100	SCI
Vydac 214 TP[a]	5, 10	30	TSG

Continued

Table 15. List of silica-based reversed-phase column support materials, *continued*

Name	Particle diameter (μm)	Mean pore diameter (nm)	Manufacturers/ suppliers
Bonded phase: C_6			
Spherisorb S5(10)X C6[a]	5 (10)	30	PSL
Bonded phase: C_8			
Apex I	3, 5, 10	10	JCL
Apex WP C8[a]	7	30	JCL
Aquapore RP-300	7	30	PRC, PEL
Astec 300A C8	5	30	BTL
Bakerbond widepore Octyl[a]	5	30	JTB
Bio-Sil	3, 5	9	BRL
Hypersil	5, 10	12, 30	SSP
ICN Silica RP 8	7–12, 18–32, 32–63	6, 10	ICN
LiChrosorb	10	–	MBH
LiChrospher CH8/II	10	50, 100, 400	MBH
Nucleosil	5, 10	10	MNG
Partisil	5, 10	8.5	WIL
Pro RPC	5	30	PMB
Sephasil C8	5, 12	12	PMB
Serva Octyl Si100[a]	5, 10	10	SFG
Spheri-5 RP-8	5, 10	8	PRC
Spherisorb[a]	3, 5, 10	8	FSL, PSL
Supelco LC-308	5	30	SUK
SynChropak RP-8[a]	6.5	30, 100	SCI

Vydac 228 TP[a]	5, 10	3	TSG
Zorbax	5	7	DPL
Bonded phase: C_{18}			
Apex I	3, 5, 10	10	JCL
Apex WP C18[a]	7	30	JCL
Astec 300A C18	5	30	BTL
Bakerbond widepore Octadecyl[a]	5	30	JTB
Bio-Sil	3, 5	8, 9	BRL
Exsil 100	3, 5	10	HCL
Hichrom RPB	5	11	HCL
Hypersil	5, 10	12, 30	SSP
ICN Silica RP 18	7–12, 18–32, 32–63	6, 10	ICN
LiChrosorb	5, 10	–	MBH
Nucleosil	3, 5, 10	10	MNG
Partisil	5, 10	8.5	WIL
Sephasil C18	5, 12	12	PMB
Serva Si100[a]	5, 10	10	SFG
Serva Si300[a]	5, 10	30	SFG
Serva Si500[a]	10	50	SFG
Spheri-5 RP-18	5	8	PRC
Spherisorb[a]	3, 5, 8, 10	8	PSL
Supelco LC-318	5	30	SUK
SynChropak RP-18[a]	6.5	10, 20, 100	SCI
TSK Gel ODS-80TM	5	8	THS
TSK Gel ODS-120A	5, 10	12	THS

Continued

Table 15. List of silica-based reversed-phase column support materials, *continued*

Name	Particle diameter (μm)	Mean pore diameter (nm)	Manufacturers/ suppliers
TSK Gel ODS-120T	5, 10	12	THS
Ultrasphere ODS	3, 5	5–9	BKI
Vydac 201 TP[a]	5, 10	30	TSG
Vydac 218 TP[a]	5, 10	30	TSG
Bonded phase: Amino			
Apex I	5	10	JCL
Hypersil	5, 10	12, 30	SSP
Nucleosil	5, 10	10	MNG
Spherisorb[a]	3, 5, 10	8	PSL
Bonded phase: Cyano			
Apex I	5	10	JCL
Apex WP CN	7	30	JCL
Bakerbond widepore Cyanopropyl[a]	5	30	JTB
Hypersil	5, 10	12, 30	SSP
Nucleosil	7	10	MNG
Spherisorb[a]	3, 5, 8, 10	8	PSL
Zorbax	5	7	DPL
Bonded phase: Phenyl			
Apex I	5	10	JCL
Apex WP Phenyl[a]	7	30	JCL
Aquapore PH-300	7	30	PRC

Astec 300A Diphenyl[a]	5	30	BTL
Bakerbond widepore Diphenyl[a]	5	30	JTB
Bio-Sil	3, 5	9	BRL
Hypersil	5, 10	12, 30	SSP
Nucleosil	5, 10	10	MNG
Spherisorb[a]	3, 5, 10	8	FSL, PSL
Supelco LC-3DP	5	30	SUK
TSK Gel Phenyl-5PN	10	–	THS
Vydac 219 TP[a]	5	30	TSG
Zorbax	5	7	DPL

[a]Bulk material available.

Table 16. List of non-silica-based reversed-phase column support materials

Name	Functional group	Particle diameter (μm)	Mean pore diameter (nm)	Manufacturers/ suppliers
Polyether based				
Biogel TSK RP +	'High-density' phenyl	10	100	BRL
TSK Phenyl 5PW	Phenyl	10	100	ACL, BRL
Polystyrene based				
Bio-Gel RP +	Phenyl	10	100	BRL
PLRPS-300A	Phenyl	10	30	PLL
PLRPS-300	C_{18} phase	8, 10	30	PLL
PLRPS-1000	C_{18} phase	8, 10	100	PLL
POROS R1[a]	C_4 phase	10, 20, 50	600–800	PSB
POROS R2[a]	C_8, C_{18} phase	10, 20, 50	600–800	PSB
SOURCE 15RPC	–	15	–	PMB

[a]Bulk material available.

Table 17. Some characteristics of commonly used matrices for affinity chromatography

Matrix type	Particle size (μm)	Measure of pore size (nm or exclusion limits)	Stability		Surface groups	Comments	Manufacturer
			pH	Max. pressure (bar)			
Natural							
Cross-linked agarose (12%)	3–10	450×10^3	2–14	–	Hydroxyl	Must not dry out	Ref. 1

Cross-linked agarose (12%)	5–40	450×10^3	2–14	–	Hydroxyl		Ref. 1
Dextran	40–125	–	2–12	–	–	Swells and shrinks depending on ionic strength	PMB
Superose 6	13	5×10^3–5×10^6	1–14	15	Hydroxyl		PMB
Superose 12	10	$< 3 \times 10^5$	1–14	30	Hydroxyl		PMB
Silica							
Hypersil WP 300	5, 10	30 nm	2–8	1035	Silanol	Unstable above pH 8	SSP
LiChrospher range	10	30–400 nm	2–8	–	Silanol		MBH
SelectiSpher-10	10	50 nm	2–8	80	Tresyl-diol-bonded		PRC
Ultraffinity-EP	–	–	2–8	–	Epoxy-bonded		BKI
Synthetic							
Affi-Prep	50	100 nm	2–12	–	*N*-hydroxy-succinimide, Protein A		BRL
Dynospheres XP-2507	20	5–200 nm	1–13	140	Hydroxyl		DPA
Polyacrylamide	–	–	2–11	–		*N-N′ bis* is toxic	BRL, LSL, PMB
Polystyrene	Up to 1100	–	0–14	–		Non-ionic interaction	BRL, SFG
Trisacryl	40–80	$> 10^7$	0–13	–	Amino, carboxyl, sulfonate		PMB
TSK-PW range	10–20	10^3–3×10^7	1–13	130	Hydroxyl		THS

Table 18. Commercially available chiral stationary support matrices

Name[a]	Class[b]	Chiral selector[c]	Manufacturer
Bakerbond Chiral Covalent DNBPG	Brush, π-acceptor	(*R*)-DNB-phenylglycine	JTB
Bakerbond Chiral Covalent DNBLeu	Brush, π-acceptor	(*S*)-DNB-leucine	JTB
Bakerbond Chiral Ionic DNBPG	Brush, π-acceptor	(*R*)-DNB-phenylglycine	JTB
Bio-Sil Chiral I	–	DNB-phenylglycine	BRL
Cellulose CEL-AC-40 XF	Helical	Cellulose triacetate	MNG
CHIRA-chrom-1	Brush, π-acceptor	DNB-phenylglycine	HCL
Chiral BDex Si100Polyol	Cavity	β-Cyclodextrin	SFG
Chiral DNBDL-C Si100Polyol	Brush, π-acceptor	(*R*)-DNB-leucine (covalent)	SFG
Chiral DNBLL-C Si100Polyol	Brush, π-acceptor	(*S*)-DNB-leucine (covalent)	SFG
Chiral DNBPG-C Si100Polyol	Brush, π-acceptor	(*R*)-DNB-phenylglycine (covalent)	SFG
Chiral-HSA	Affinity	Human serum albumin	HCL
Chiral HyproCu Si100Polyol	Ligand exchange	Hydroxyproline-copper	SFG
Chiral ProCu Si100Polyol	Ligand exchange	Proline-copper	SFG
Chiral Protein BSA	Affinity	Bovine serum albumin	LSI
Chiral Protein HSA	Affinity	Human serum albumin	LSI
Chiral ValCu Si100Polyol	Ligand exchange	Valine-copper	SFG
ChiraSpher	Helical	Poly(*N*-acryloylphenyalanine ethyl ester)	MBH
Covalent D-Phenyl Glycine	Brush, π-acceptor	(*R*)-DNB-phenylglycine	RTI
Covalent L-Phenyl Glycine	Brush, π-acceptor	(*S*)-DNB-phenylglycine	RTI
Covalent D,L-Phenyl Glycine	Brush, π-acceptor	(*R*, *S*)-DNB-phenylglycine	RTI
Covalent L-Leucine	Brush, π-acceptor	(*S*)-DNB-leucine	RTI
EnantioPac	Affinity	α_1-Acid glycoprotein	PMB
Ionic D-Phenyl Glycine (Pirkle Type 1A)	Brush, π-acceptor	(*R*)-DNB-phenylglycine	RTI

Ionic L-Leucine	Brush, π-acceptor	(S)-DNB-leucine	RTI
Nucleosil Chiral-1	Ligand exchange	Hydroxyproline-copper	MNG
Nucleosil Chiral-2	Brush	DNB-phenylglycine	MNG
Nucleosil Chiral-3	Brush	DNB-phenylglycine	MNG
Pirkle – first series	π-acceptor	DNB-phenylglycine	RTI
Pirkle – second series	π-donator		RTI
Resolvosil	Affinity	Bovine serum albumin	MNG
Supelcosil LC-(R)-Urea	Brush, urea type	Phenylethylurea	SUK
TSK Gel Enantio-Li	Cavity	Chiral polymer	THS
TSK Gel Enantio-Si	Cavity	Chiral polymer	THS

[a] All matrices are based on silica with the exception of Cellulose CEL-AC-40 XF.
[b] Brush type phases always include dipole and hydrogen bonding interactions.
[c] DNB, 3, 5-dinitrobenzoyl.

Chapter 4 SAMPLE PREPARATION

1 Introduction

This chapter does not seek to duplicate the detailed methods for the preparation of molecules offered in many manuals; the reader should consult appropriate texts for such information [1–8]. However, the aim of this chapter is to provide a reference source for the different types of reagents and methods used for the extraction, derivatization and labeling of both macromolecules and small molecules. Also included are standards used in various chromatography systems. The following tables are not intended to be exhaustive and there are recipes and referenced techniques other than those listed. Suppliers have been given in abbreviated form; consult Chapter 8 for the complete names and addresses.

(*Tables 1–10*). Where appropriate, reagents for purification and concentration by precipitation are also included. However, owing to space limitations, assays for determining the concentration of specific molecules are not included; it is advised that the reader consult appropriate manuals.

3 Derivatization and labeling methods

The characteristics of many native molecules may not permit direct estimation in solution and so, to facilitate detection and quantification of particular molecules, it is often necessary to derivatize the sample components to a modified form which then lends itself to detection. A selection of derivatization techniques for various molecules are listed in *Tables 11–17*; see also Chapter 6, Section 3 for reagents for detection. For small samples it is convenient to radiolabel

2 Extraction methods

The successful application of liquid chromatography often depends upon the sample preparation procedure used. In limited instances, direct injection of biological fluids, such as bile, serum and urine, is possible. However, it is common if not essential practice to remove from the initial sample the large quantities of interfering or contaminating materials that are present. This results in better separation, increased resolution and faster analysis. A selection of reagents and methods used for the extraction of molecules such as proteins, nucleic acids, lipids, and many more is presented

the molecules to enhance detection; reagents for isotopic labeling are given in *Tables 18–20*.

4 Molecular weight markers and standards

Molecular weight markers or standards are used to calibrate the separation system. They are used for measurements of capacity, mass and activity recoveries. In addition, it is common practice to optimize chromatographic parameters to maximize resolution, recovery and throughput using the standards. *Tables 21–24* list standards for various separations. Commercial sources of standards are given in *Table 25*.

Table 1. Intracellular component and protein release techniques [1, 9]

Technique	Applications	Additional information
Disruption by physical methods		
Abrasives	Yeast cells	Alumina or sand used as abrasives
Homogenization	Animal tissue, cell suspensions	Widely used technique
Liquid extrusion	Cell suspensions/pastes	Several passes at high speed through device are required to achieve complete release of intracellular components
Mixers and blenders	Plant tissue, fungal mycelium	Not satisfactory for the disruption of more robust microbial cells
Solid extrusion	Cell pastes	Unsuitable for preparing proteins that are damaged by multiple passes
Ultrasonication	Microbial cells	Heat produced as a result of sonication may cause protein denaturation, and so cell paste should be kept on ice and sonication limited to short bursts
Nonphysical methods of disruption		
Detergents and solvents	Bacteria, animal cells	Protein denaturation and precipitation may limit use
Freeze–thaw	Susceptible microbes and eukaryote cells	Not widely used as many organisms are resistant to rupture
Lytic enzymes	Bacteria, yeast cells	Particularly applicable where organisms are resistant to mechanical disruption, or contain sensitive products
Osmotic shock	Animal and plant cells, some Gram-negative bacteria	The resultant shock is often insufficient to burst the cell, but it allows some proteins to be released from the cells

Table 2. Chart for ammonium sulfate precipitation of proteins

Initial concn of ammonium sulfate, % saturation at 0°C	Final concentration of ammonium sulfate, % saturation at 0°C (g solid ammonium sulfate to add to 100 ml of solution)																
	20	25	30	35	40	45	50	55	60	65	70	75	80	85	90	95	100
0	10.7	13.6	16.6	19.7	22.9	26.2	29.5	33.1	36.6	40.4	44.2	48.3	52.3	56.7	61.1	65.9	70.7
5	8.0	10.9	13.9	16.8	20.0	23.2	26.6	30.0	33.6	37.3	41.1	45.0	49.1	53.3	57.8	62.4	67.1
10	5.4	8.2	11.1	14.1	17.1	20.3	23.6	27.0	30.5	34.2	37.9	41.8	45.8	50.0	54.4	58.9	63.6
15	2.6	5.5	8.3	11.3	14.3	17.4	20.7	24.0	27.5	31.0	34.8	38.6	42.6	46.6	51.0	55.5	60.0
20	0	2.7	5.6	8.4	11.5	14.5	17.7	21.0	24.4	28.0	31.6	35.4	39.2	43.3	47.5	51.9	56.5
25		0	2.7	5.7	8.5	11.7	14.8	18.2	21.4	24.8	28.4	32.1	36.0	40.2	44.2	48.5	52.9
30			0	2.8	5.7	8.7	11.9	15.0	18.4	21.7	25.3	28.9	32.8	36.7	40.8	45.1	49.5
35				0	2.8	5.8	8.8	12.0	15.3	18.7	22.1	25.8	29.5	33.4	37.4	41.6	45.9
40					0	2.9	5.9	9.0	12.2	15.5	19.0	22.5	26.2	30.0	34.0	38.1	42.4
45						0	2.9	6.0	9.1	12.5	15.8	19.3	22.9	26.7	30.6	34.7	38.8
50							0	3.0	6.1	9.3	12.7	16.1	19.7	23.3	27.2	31.2	35.3
55								0	3.0	6.2	9.4	12.9	16.3	20.0	23.8	27.7	31.7
60									0	3.1	6.3	9.6	13.1	16.6	20.4	24.2	28.3
65										0	3.1	6.4	9.8	13.4	17.0	20.8	24.7
70											0	3.2	6.6	10.0	13.6	17.3	21.2
75												0	3.2	6.7	10.2	13.9	17.6
80													0	3.3	6.8	10.4	14.1
85														0	3.4	6.9	10.6
90															0	3.4	7.1
95																0	3.5

Table 3. Extraction and precipitation methods for nucleic acids

Nucleic acid	Reagents	Comments	Ref.
DNA	Lysis solution (10 mM Tris–HCl pH 8.0, 0.5% SDS, 0.1 M EDTA, 20 μg ml^{-1} pancreatic RNase); Tris-buffered saline (25 mM Tris base, 0.2 M NaCl, 2.5 mM KCl, 15 μg ml^{-1} phenol red pH 7.4, autoclaved); TE (10 mM Tris, 1 mM EDTA); proteinase K (20 mg ml^{-1}); buffer-saturated phenol (pH 8.0); 10 M ammonium acetate (filtered); ethanol (absolute and 70%)	Is used to isolate genomic DNA from mammalian and plant cells. However, this method is prone to shearing of the very long DNA molecules during isolation	11
RNA (phenol-chloroform-isoamyl alcohol method)	Lysis buffer[a]; buffer-saturated phenol-chloroform-isoamyl alcohol (25:24:1); absolute and 70% ethanol; isopropanol; 3 M sodium acetate pH 5.2; sterile, double-distilled water treated with 0.1% (w/v) DEPC; chloroform-isoamyl alcohol (24:1)	A major drawback with phenol extraction is that, depending on the nature of the starting material, the RNA can be contaminated with significant amounts of both DNA and polysaccharides as both can be present in the aqueous layer after phenol extraction. Both contaminants are precipitated along with the RNA by ethanol or isopropanol and, depending on the subsequent use of the RNA, they may affect further analyses. Has been largely superseded by the acid guanidinium thiocyanate method	12

RNA (acid guanidinium thiocyanate method)	GTC solution (4 M guanidinium thiocyanate, 25 mM sodium citrate pH 7.0, 0.5% sarcosyl, 0.1 M 2-mercaptoethanol); buffer-saturated phenol-chloroform-isoamyl alcohol (25:24:1); ethanol (absolute and 70%); isopropanol; 3 M sodium acetate pH 5.2; sterile, double-distilled water treated with 0.1% (w/v) DEPC; chloroform-isoamyl alcohol	This method uses the chaotropic agent guanidinium thiocyanate in which the cation and the anion are both strongly chaotropic. This allows a very fast inactivation of cellular ribonucleases (RNases) upon cell disruption. A fast inactivation is often essential making this the method of choice. The tissue types used with this method are highly diverse; bacterial, plant and animal tissue have all successfully had RNA extracted from them	13

[a]See ref. 14 for a review of lysis buffers.

DEPC, diethylpyrocarbonate; EDTA, ethylenediamine tetraacetic acid; SDS, sodium dodecyl sulfate.

Table 4. Extraction of amino acids and small peptides and derivatization to OPT-amino acids

Preparation	Method	Comments
Extraction	Homogenize tissue in 4 vols of methanol. Allow to stand for 10 min at 4°C. Centrifuge at 5000–10 000 *g* for 10 min to remove precipitated protein	If acid is needed to efficiently disrupt the tissue and precipitate protein, the acid will need to be neutralized as the OPT reaction is performed at alkaline pH. Orthophosphoric acid is preferred as it will not interfere with the separation after neutralization
Derivatization	To 50 μl of supernatant add 200 μl of OPT reagent, mix and allow to stand for 2 min. Inject directly into HPLC column	

OPT, *o*-phthalaldehyde-thiol.

Table 5. Preparation of glycopeptides and oligosaccharides for affinity chromatography

Method	Extraction solutions	Comments	Ref.
Alkaline borohydride elimination	0.1 M NaOH, 1 mM $NaBH_4$	Breaks *O*-glycosidase linkages between carbohydrate chains and protein	15
Peptide: *N*-glycosidase F (PNGase F)	1% SDS; 0.25 M sodium phosphate pH 6.8, containing 0.6% Nonidet P-40; PNGase F	Hydrolyzes all types of oligosaccharide that are linked to protein through a GlnNAc-Asn, *N*-glycosidic bond	16
Pronase digestion	0.1 M Tris-acetate buffer pH 7.8, containing 0.15 M NaCl, 1 mM calcium acetate; pronase	Toluene is included to inhibit microbial growth	15

SDS, sodium dodecyl sulfate.

Table 6. Lipid extraction methods [17]

Lipid	Reagents	Comments
Eicosanoids	100% or 30% methanol in water; ethanol	The use of 30% methanol in water gives better recovery and dissolves any salts present
Eicosanoids	Methanol or ethanol; citric acid or formic acid; sodium sulfate; chloroform; methanol:water (30:70) or desired solvent	Extraction with chloroform:methanol (1:1) will give a purer fraction which is more suitable for derivatization procedures
Fatty acids	Methanol:chloroform:heptane (1.41:1.25:100, by vol.); 0.1 M sodium carbonate buffer pH 10.5	System devised for extraction of product fatty acids [18]

Glycolipids	Chloroform:methanol (2:1, v/v), (1:1, v/v), (1:2, v/v); water	System devised by Folch *et al.* [19]. Extraction with higher methanol concentrations can be replaced by treatment with 0.4 M sodium acetate in chloroform:methanol:water (30:15:4, by vol.; 1 g, 40 ml) at room temperature for 24 h [20]
Phospholipids	Chloroform:methanol (2:1); solution of the chloride of one of the following cations: Na^+, K^+ (0.05 M), Ca^{2+}, Mg^{2+} (0.0015 M); chloroform:methanol:water (3:48:47); desired solvent	System devised by Folch *et al.* [19]. Further purification of phospholipid extract can be achieved using precipitation with ice-cold acetone [21]
Triglycerides	Chloroform:methanol (2:1)	Chloroform:methanol (2:1) will extract all neutral lipids from tissues or biological fluids; however, some free fatty acids and phospholipids will also be extracted. It is necessary to purify this fraction using columns packed with bonded silica adsorbent

Table 7. Extraction of steroids

	Reagents	Ref.
Bile acids	10% $ZnSO_4$; methanol	22
	Ethanol; methanol	23
Steroids	*n*-hexane/isopropyl alcohol (3:2, v/v); Lipidex 1000; Lipidex 5000	24

Table 8. Vitamin extraction methods [25]

Vitamin	Source	Reagents	Ref.
Provitamin A carotenoids	Plasma	Echinenone in ethanol; *n*-hexane; desired solvent	26
Retinoic acid	Plasma	All-*trans*-13-dimethyl retinoic acid in ethanol; ethanol; 2 M NaOH; *n*-hexane; 2 M HCl; desired solvent	27
Retinol	Plasma	All-*trans*-retinyl acetate in ethanol; *n*-hexane; ethanol	
Retinyl esters	Plasma	Water; retinyl propionate in methanol; chloroform; methanol: chloroform (4:1, v/v)	28
Riboflavin (Vitamin B_2)	Urine	Oxalic acid dihydrate	29
Thiamin (Vitamin B_1)	Whole blood	10% (w/v) perchloric acid; 15 μM salicylamide, 0.6 M NaOH, 1.8 M sodium acetate solution; acid phosphatase solution (10 mg ml^{-1} in 0.9% w/v NaCl)	30
Vitamin C	Whole blood	0.3 M trichloroacetic acid; 4.5 M sodium acetate buffer pH 6.2; ascorbic acid oxidase spatula; 0.1% (w/v) solution of *o*-phenylenediamine	31
Vitamin D and metabolites	Plasma	Methanol:chloroform (2:1, v/v); chloroform; water	

Table 9. Extraction of nucleotides [32]

Source	Method
Whole tissue	Liquid nitrogen ($-196^{\circ}C$); 10% (w/v) TCA; water-saturated diethyl ether
Red blood cells	10% (w/v) TCA; water-saturated diethyl ether
Platelets and other white cells	10% (w/v) TCA; water-saturated diethyl ether

TCA, trichloroacetic acid.

Table 10. Porphyrin extraction methods

Scheme	Source	Reagents	Ref(s)
Acetone–water extraction	Red blood cells	Celite (5 g l^{-1} in isotonic saline with detergent); acetone:water (4:1, v/v)	33
Ethanol extraction	Red blood cells	Water; 95% ethanol	34
Ether–acetic acid extraction	Plasma	Ether:acetic acid (4:1, v/v); 2.7 M HCl	35, 36
Ethyl acetate–acetic acid extraction	Red blood cells	Ethyl acetate:acetic acid (3:1, v/v); mesoporphyrin	37

Table 11. Protein modification; recognition sequences and donors

Modification	Recognition sequence	Group donor	Ref(s)
Acetylation	Amino terminal residues	Acetyl CoA	38
ADP ribosylation	Glu, Arg	NAD^+	39, 40
Amidation	Carboxy terminal Gly	Conversion of terminal Gly	41
Carboxylation	Glu-X-X-X-Glu-X-Cys	CO_2	42, 43
Glycosylation	Asn-X-Thr/Ser	Sugar nucleotide diphosphates	44, 45
Hydroxylation		Oxygen	46
3-Pro	X-Pro-Gly		
4-Pro	Pro-(4-Hyd)Pro-Gly		
Isoprenylation	Cys-Ali-Ali-X-OOH	Mevalonyl CoA	47, 48
Methylation		*S*-Adenosyl methionine	49
Myristoylation	NH_2 terminal Gly	Myristoyl CoA	50, 51
Phosphorylation	Many	ATP	52, 53
Sulfonation	Acidic residues	PAPS	54

Ali, aliphatic residue; PAPS, adenosine-3′-phospho-5′-phosphosulfate.

Table 12. Chemical modification of amino groups of proteins

Reagent	Ref(s)
Acylation	
Acetic anhydride	55
Acetylimidazole	56, 57
N-Acetylsuccinimide	56
N-Hydroxysuccinimide acetate	58
Maleic anhydride	59
Succinic anhydride	60
Alkylation and arylation	
1-Fluoro-2,4-dinitrobenzene	61, 62
Iodoacetic acid	63
2,4,6-Trinitrobenzenesulfonic acid	64
Amidination	
Methyl acetimidate	65-67
Carbamylation	
Potassium cyanate	68
Guanidination	
1-Guanyl-3,5-dimethylpyrazole nitrate	69
O-Methylisourea	70

Table 13. Hydrolysis techniques used for carbohydrate analysis

	Incubation conditions	
Method	Time (h)	Temperature (°C)
1–2 M HCl	6	100
0.01 M HCl containing Dowex 50X2 resin in the H^+ form	48	95
0.1–0.5 M H_2SO_4	8–16	100
0.25 M H_2SO_4 in 90% acetic acid	4–16	80
2 M trifluoroacetic acid	1.5	120

Table 14. Derivatization techniques used after hydrolysis for carbohydrate analysis [71]

Technique	Reagents
Deamination of de-*N*-acetylated monosaccharides	Solid sodium nitrite
De-*O*-acetylation	Dry methanol; barium or sodium methoxide or KOH pellets dissolved in methanol: toluene (3:1, v/v) to a final concentration of 0.1 M
Re-*N*-acetylation	Acetic anhydride; saturated sodium bicarbonate
O- and *N*-acetylation	Acetic anhydride; dry pyridine

Table 15. Methanolysis and subsequent derivatization techniques used for carbohydrate analysis

Reaction step	Procedure
Methanolysis	Dissolve sample in dry methanol containing 1 M HCl. Heat for 16 h at 80°C
Neutralization	Add powdered silver carbonate. Centrifuge to remove the precipitated silver salt, and collect supernatant
N-Acetylation	Add acetic anhydride to a final concentration of 10% acetic anhydride

Table 16. Pre-column derivatization of carbohydrates for HPLC

Carbohydrate	Reagent(s)	Ref(s)
Amino sugars	*o*-Phthalaldehyde	72
Glycoprotein hexoses and hexosamines	Phenylisothiocyanate	73
Monosaccharides	Benzoyl chloride	74
Oligosaccharides	Aminopyridine	75, 76
	Ethyl 4-aminobenzoate	77, 78
	1-(4-methoxy)phenyl-3-methyl-5-pyrazolone	79
Reducing sugars	Aminopyridine	80
	Dansyl hydrazine	81
Sialic acids	1,2-Diamino-4,5-methylenedioxybenzene	82

Table 17. Derivatization of lipids

Derivatization	Reagents	Comments	Ref.
Methyl esters of fatty acids	Boron trifluoride-methanol (14% w/v); hexane; saturated NaCl solution; $NaSO_4$; appropriate solvent, e.g. acetone or acetonitrile	Boron trichloride or anhydrous HCl can be used in place of boron trifluoride; alternatively, methyl esters of fatty acids can be prepared using dimethylformamide (DMF)-dimethylacetal	17
Fatty acid methyl esters by transesterification of lipids	Sodium-dried ether; dry methyl acetate; 1 M sodium methoxide in dry methanol; saturated oxalic acid in ether; appropriate solvent, e.g. fresh ether or hexane		83
Fatty acid methyl esters by transesterification of lipids	Acetonitrile; methyl acetate; 1 M potassium methoxide in methanol; acetic acid	This method is more compatible with reversed-phase HPLC	84
4-methyl-7-methylcoumarin fatty acid esters	10% KOH in methanol; crown ether solution; 4-bromomethyl-7-methoxycoumarin (BMC)	Fluorescent derivative used for enhanced detection; solid potassium carbonate catalyst may be used in place of KOH	85
p-Bromophenacyl bromide (pBPB) eicosanoids	Acetonitrile; pBPB; diisopropylethylamine	BMC may be used in place of pBPB	17

Table 18. Reagents for the isotopic labeling of proteins

Reacting group	Labeling reagents
-CH_2OH Aliphatic hydroxyl groups (serine/threonine residues)	Acetic anhydride; diisopropylphosphofluoridate
-NH_2 Free amino groups (N-terminal or lysine residues)	Acetic anhydride; Bolton and Hunter reagent; dansyl chloride; ethyl acetimidate; 1-fluoro-2,4-dinitrobenzene; formaldehyde; isethionyl acetimidate; maleic anhydride; methyl 3,5-diiodohydroxybenzimidate; phenyl isothiocyanate; potassium borohydride; sodium borohydride; succinic anhydride; *N*-succinimidyl propionate
N NH Imidazole groups (histidine residues)	Dansyl chloride; iodine
OH Phenolic hydroxyl groups (tyrosine residues)	Acetic acid; dansyl chloride; iodine
-SH Thiol groups (cysteine residues)	Acetic acid; bromoacetic acid; chloroacetic acid; *p*-chloromercuribenzenesulfonic acid; *p*-chloromercuribenzoic acid; dansyl chloride; *N*-ethylmaleimide; iodoacetamide; iodoacetic acid

Table 19. Chemical and enzymatic techniques for radiolabeling glycopeptides [14]

	Mechanism	Reagents	Comments
^{14}C-acetylation of glycopeptides		^{14}C-labeled acetic acid anhydride (5–10 mCi $mmol^{-1}$); unlabeled acetic acid anhydride; 0.5% (w/v) $NaHCO_3$; water	Radiolabeled acetic acid anhydride should be handled in a fume cupboard
Galactose	Galactose residues and GalNAc units are labeled using galactose oxidase/^3H-$NaBH_4$. The enzyme attacks end-chain galactose	Glycopeptides; galactose oxidase 12 U ml^{-1}); horseradish peroxidase (18 U ml^{-1}); 20 mM sodium phosphate, 0.15 M NaCl, 0.45 M sodium acetate, pH 7.0 solution; 100% trichloroacetic acid; water; 50 mM sodium phosphate, 0.15 M NaCl pH 7.5 solution; ^3H-$NaBH_4$ (200–300 mCi $mmol^{-1}$) in 10 mM NaOH; unlabeled 0.1 M $NaBH_4$ in 10 mM NaOH	Excess $NaBH_4$ can be removed from the [^3H]glycopeptides by dialysis or gel filtration. If labeling is poor, treat glycopeptides with neuraminidase prior to labeling. Experiments should be performed in a fume cupboard
Sialic acid	Hydroxyl groups in the acyclic carbons (C7–C9) of sialic acid may be specifically oxidized by dilute sodium periodate ($NaIO_4$) to yield a seven-carbon sugar with a C7 aldehyde which can then be reduced with ^3H-$NaBH_4$	Glycopeptides containing 1 mol of sialic acid; 0.1 M sodium acetate, 0.15 M NaCl pH 5.6 solution; 10 mM $NaIO_4$; glycerol; 50 mM sodium phosphate, 0.15 M NaCl pH 7.5 solution; ^3H-$NaBH_4$ in 10 mM NaOH	Excess $NaBH_4$ can be removed from the [^3H]glycopeptides by dialysis or gel filtration. Experiments should be performed in a fume cupboard

Table 20. Radioactive labeling methods of nucleotides

Radionuclide	Specific activity of labeled nucleotides (TBq $mmol^{-1}$)	Labeling methods	Specific activity of probe (dpm μg^{-1}) ($\times 10^{8}$)
^{3}H	0.9–3.7	Nick-translation	0.5
		Random-priming	1.5
		Phage RNA polymerase	0.5
^{32}P	14.8–222.2	Nick-translation	5
		Random-priming	50
		Phage RNA polymerase	13
		End-labeling	0.05
^{33}P	14.8–111.0	Nick-translation	1
		Random-priming	20
^{35}S	14.8–55.5	Nick-translation	1
		Random-priming	7
		Phage RNA polymerase	13
^{125}I	37.0–74.0	Nick-translation	1
		Random-priming	15
		Phage RNA polymerase	10
		Direct iodination	2

Table 21. Protein standards for size exclusion HPLC

Protein	Molecular weight (kDa)
Blue Dextran 2000	2000
Thyroglobulin	660
Apoferritin	443
Catalase	250
Aldolase	158
Bovine serum albumin	67
Ovalbumin	40
Chymotrypsinogen A	25
Ribonuclease A	14
Cytochrome *c*	13
Aprotinin	6.5

Table 23. Standards used for nucleotides, nucleosides and bases

Nucleotides	Bases/nucleosides
CMP	Theophylline
XMP	Cyclic nucleotides
XDP	
Polynucleotides	
Cyclic nucleotides	
Radiolabeled bases	Radiolabeled bases

Table 24. Internal standards used for HPLC of porphyrins

Methyl ester	Concentration (nmol l^{-1})
Coproporphyrin I	170
Protoporphyrin IX	215
Uroporphyrin I	235

Table 22. Protein standards for ion-exchange HPLC

Protein	Molecular weight (kDa)	Isoelectric point (pI)
Thyroglobulin	660	4.6
Urease	480	5.1
Fibrinogen	330	5.5
Catalase	250	5.6
Immunoglobulins G	150	6.4–7.2
Lactic acid dehydrogenase	140	–
Bovine serum albumin	67	–
Human serum albumin	66.5	4.8
Hemoglobin, horse	65	6.9
Ovalbumin	40	4.6
Pepsin	35.5	< 1.0
Carboxypeptidase	34	6.0
Carbonic anhydrase	30	7.3
Soybean trypsin inhibitor	22.5	–
Human growth hormone	21.5	6.9
Horse myoglobin	17	7.0
Hen egg white lysozyme	14.3	11.0
Ribonuclease	14	7.8
Cytochrome *c*	13	10.6
Insulin	5.7	5.3

Table 25. Commercial sources of molecular weight markers and standards

Molecular weight markers/standards	Supplier(s)
Size exclusion	
Protein	BRL, PMB, SCC
Hydrophobic interaction	
Protein	BRL
Ion-exchange	
Protein	BRL
Reversed-phase	
Nucleotide	SCC
Peptide	PRC, SCC
Protein	SCC

Chapter 5 ADSORBENTS AND SOLVENT SYSTEMS

1 Introduction

The following tables list a selection of the different types of solvent system (mobile phases and eluants) used for liquid chromatography of both small and macromolecules. Also included is a section on adsorbents used for affinity chromatography. The tables presented are not intended to be exhaustive and there are recipes and referenced techniques other than those listed.

2 Adsorbents and immobilized ligands used in affinity chromatography

It is sometimes necessary to prepare an affinity adsorbent for a specific purpose. This can be achieved by reacting a chosen ligand either with an activated derivative of a solid support, (e.g. the introduction of hydroxyl or amino groups) or by matrix activation (e.g. the introduction of epoxy or isothiocyanato groups), depending on the organosilane used. Functional matrices react directly only with active ligands, for example, reactive dyes, whereas activated silica reacts directly with ligands carrying amino, hydroxyl or sulfhydryl groups. *Tables 3* and *4* list reagents used for silica gel coating, functionalization and activation.

Tables 5 and *6* list various ligands immobilized to support matrices, while *Table 7* summarizes methods employed to determine the concentration of ligand bound to the support matrix.

The reader should bear in mind that the correct choice of coupling gel is dictated both by the types of group available on the ligand molecule for coupling and by the nature of the or with a derivatized support carrying a spacer which terminates in a reactive group. *Tables 1* and *2* list a selection of activated derivatives of agarose and agaroses with spacers attached; using such matrices as starting materials, the attachment of the ligand to complete the affinity adsorbent is readily achieved. Other solid supports may be used in place of agarose. Activation and coupling procedures employed for preparation of adsorbents of particular specificity have been omitted; the procedures have been reviewed elsewhere [1–3].

Silica gels must first be coated with an appropriate organosilane before ligand coupling. This compulsory coating process is accompanied by either matrix functionalization binding reaction with the substance to be purified. Immobilization should be attempted through the least critical region of the molecule to ensure minimal interference with the normal binding reaction.

3 Solvent systems and additives used in various liquid chromatographic techniques

A selection of mobile phases and eluants have been listed for various applications (*Tables 8–33*). Included are buffer substances, salts, and additives necessary for liquid chromatography of molecules.

Table 1. A selection of activated agaroses

Type of activation	Reactive group of activated gel	Reactive group of ligand	Mode of coupling of ligand to the gel	Additional information
Carbonyl-diimidazole-activated	—O—C(=O)—N(imidazole)	R-NH_2	—O—C(=O)—NH—R	Alternative to CNBr activation for coupling proteins directly to agarose. Coupled product is more stable than with CNBr activation, so reducing leakage of ligand, and substituted carbamate coupling group is uncharged. Imidazolylcarbamate group of activated gel is only slowly hydrolyzed at mildly alkaline pHs
Cyanogen bromide (CNBr)-activated	—O—C≡N (—O)₂C=NH	R-NH_2	—O—C(=NH)—NH—R (—O)₂C=N—R —O—C(=O)—NH—R	Most widely used method for coupling proteins and nucleic acids directly to agarose. Prepared by reaction of CNBr with agarose. In CNBr-activated agarose, the cyanate ester groups are reactive but unstable under alkaline conditions (stable below pH 4), whereas cyclic imidocarbonate groups are stable to alkali but less reactive. In the coupled product, the *N*-substituted isourea groups are not completely stable. The isourea group also carries a positive charge ($pK_a \approx 9.5$), giving the adsorbent ion-exchange properties

N-Hydroxy-succinimide-activated	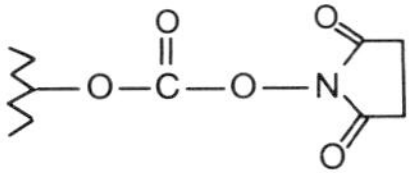 $\sim O{-}C({=}O){-}O{-}N$(succinimide)	$R{-}NH_2$ ($R{-}SH$)	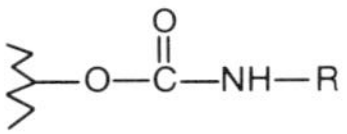 $\sim O{-}C({=}O){-}NH{-}R$	Prepared by reaction of *N*-hydroxysuccinimidochloroformate with agarose. The *N*-hydroxysuccinimide ester bond in the activated gel is rapidly hydrolyzed above pH 8.5. Coupled product is more stable than with CNBr activation and the coupling group is uncharged
p-Nitrophenyl-activated	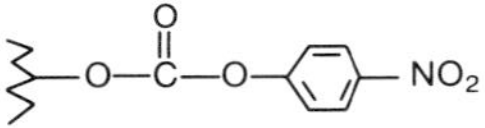 $\sim O{-}C({=}O){-}O{-}C_6H_4{-}NO_2$	$R{-}NH_2$	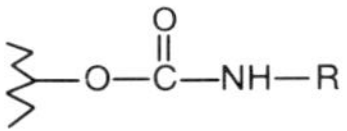 $\sim O{-}C({=}O){-}NH{-}R$	Prepared by reaction of *p*-nitrophenyl chloroformate with agarose. Coupled product is more stable than with CNBr activation and the coupling group is uncharged
Tresyl-activated	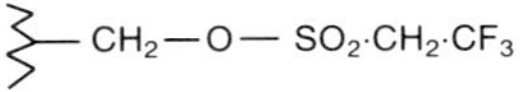 $\sim CH_2{-}O{-}SO_2{\cdot}CH_2{\cdot}CF_3$	$R{-}NH_2$	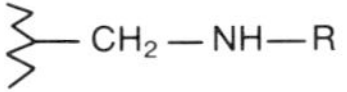 $\sim CH_2{-}NH{-}R$	Prepared by reaction of 2,2,2-trifluoroethanesulfonyl chloride (tresyl chloride) with agarose. Activated gel reacts with sulfhydryl groups, as well as amino groups, forming stable bonds
		$R{-}SH$	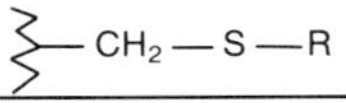 $\sim CH_2{-}S{-}R$	

Adapted from Dawson, R.M.C. *et al.* (1991) *Data for Biochemical Research*, Oxford University Press, Oxford.

Table 2. Selection of agarose matrices with spacers attached

Type	Structure	Reactive group of ligand	Mode of coupling of ligand	Additional information
Spacer terminating in amino group $-NH_2$	$\wr-NH\cdot(CH_2)_n\cdot NH_2$	R-COOH	—NH—CO—R	Spacer can be extended by reaction with succinic anhydride. Coupling is usually carried out with a water-soluble carbodiimide, such as 1-ethyl-3-(3-dimethylaminopropyl) carbodiimide (EDAC)
	$\wr-O\cdot CH_2\cdot CO\cdot NH\cdot(CH_2)_2\cdot NH_2$	R-COOH	—NH—CO—R	Six-atom hydrophilic spacer
	$\wr-CH_2\cdot NH\cdot(CH_2)_3\cdot NH\cdot(CH_2)_3\cdot NH_2$	R-COOH	—NH—CO—R	Nine-atom spacer
	$\wr-O\cdot CH_2\cdot CHOH\cdot CH_2\cdot O\cdot(CH_2)_4-O-CH_2\cdot CHOH\cdot CH_2\cdot NH\cdot(CH_2)_n\cdot NH_2$	R-COOH	—NH—CO—R	Diaminoalkane is attached to epoxy-activated agarose, giving a spacer containing 13 + *n* atoms
Spacer terminating in carboxyl group -COOH	$\wr-O\cdot CH_2\cdot COOH$	$R-NH_2$	$\wr-O\cdot CH_2\cdot CO-NH-R$	Coupling is usually carried out with a water-soluble carbodiimide, such as EDAC

	$\sim NH \cdot (CH_2)_5 \cdot COOH$	$R\text{-}NH_2$	$\sim NH \cdot (CH_2)_5 \cdot CO—NH—R$	
	$\sim O \cdot CH_2 \cdot CO \cdot NH \cdot (CH_2)_2 \cdot NH$ ∣ $HOOC—(CH_2)_2 \cdot CO$	$R\text{-}NH_2$	$—CO—NH—R$	Hydrophilic 10-atom spacer
Spacer terminating in epoxy group $—CH—CH_2$ (epoxide, O) (epoxy-activated agarose)	$\sim O \cdot CH_2 \cdot CHOH \cdot CH_2 \cdot O$ ∣ $(CH_2)_4$ ∣ $CH_2—CH—CH_2—O$ (epoxide, O)	$R\text{-}NH_2$ $R\text{-}OH$ $R\text{-}SH$	$—CH—CH_2$ (epoxide, O) → $—CHOH—CH_2—NH—R$ $—CHOH—CH_2—O—R$ $—CHOH—CH_2—S—R$	Prepared by reaction of agarose with the bisoxirane, 1,4 bis (2,3-propoxy)-butane (1,4-butanediol diglycidyl ether). Forms a 12-atom spacer on coupling with ligand. Coupling with ligand requires high pHs, e.g. pH 12–13, but spacer is not cleaved from agarose under these conditions. Couples with amino, hydroxyl and sulfhydryl groups forming stable secondary amine, ether and thioether bonds, respectively. Useful for ability to couple to hydroxyl-containing ligands, e.g. carbohydrates
Spacer terminating in hydrazide group $\text{-}NH.NH_2$	$\sim NH \cdot NH \cdot CO \cdot (CH_2)_4 \cdot CO \cdot NH \cdot NH_2$	$R\text{-}CHO$	$—NH—N{=}CH—R$ $\downarrow NaBH_4$ $—NH—NH—CH_2—R$	Useful for coupling with aldehyde groups, e.g. those formed in carbohydrates and nucleotides by periodate oxidation

Continued

Table 2. Selection of agarose matrices with spacers attached, *continued*

Type	Structure	Reactive group of ligand	Mode of coupling of ligand	Additional information
Spacer terminating in *N*-hydroxysuccinimide ester (*N*-hydroxysuccinimide-activated agarose)	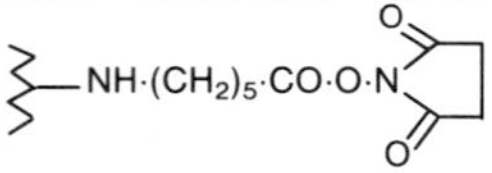 $\sim NH\cdot(CH_2)_5\cdot CO\cdot O\cdot N$ (succinimide ring)	R-NH_2	—CO—NH—R	Forms six-atom hydrophobic spacer. *N*-Hydroxysuccinimide ester is rapidly hydrolyzed at pHs above 8.5
	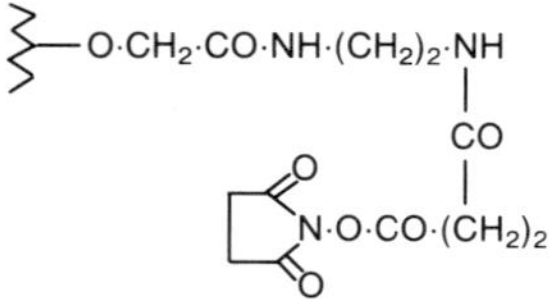 $\sim O\cdot CH_2\cdot CO\cdot NH\cdot(CH_2)_2\cdot NH$—CO—$(CH_2)_2\cdot CO\cdot O\cdot N$ (succinimide ring)	R-NH_2	—CO—NH—R	Hydrophilic, uncharged 10-atom spacer
	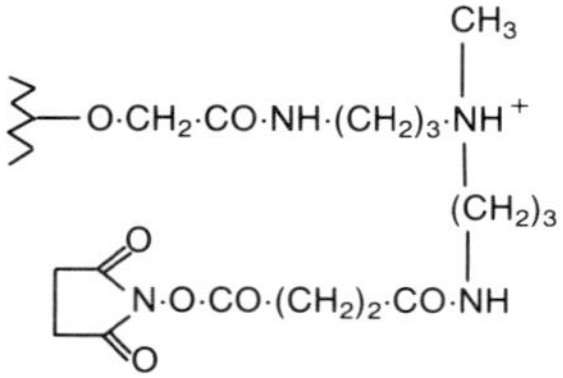 $\sim O\cdot CH_2\cdot CO\cdot NH\cdot(CH_2)_3\cdot NH^+(CH_3)$—$(CH_2)_3$—$NH\cdot CO\cdot(CH_2)_2\cdot CO\cdot O\cdot N$ (succinimide ring)	R-NH_2	—CO—NH—R	Hydrophilic, cationic 15-atom spacer

Spacer terminating in thiol group -SH	$\sim NH \cdot CH_2 \cdot CH_2 \cdot SH$	R-SH	—S—S—R	Reversible coupling of ligand by disulfide bond
	$\sim NH \cdot (CH_2)_2 \cdot NH \cdot CO \cdot CH_2 \cdot CH_2 \cdot SH$	R-SH	—S—S—R	
	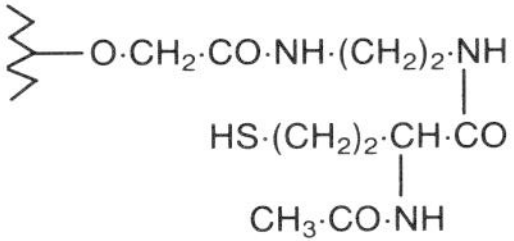 $\sim O \cdot CH_2 \cdot CO \cdot NH \cdot (CH_2)_2 \cdot NH$ — $CO \cdot CH(CH_2)_2 \cdot SH$ (with $CH_3 \cdot CO \cdot NH$ on CH), i.e. $HS \cdot (CH_2)_2 \cdot CH(NH \cdot CO \cdot CH_3) \cdot CO$	R-SH	—S—S—R	Formed by coupling *N*-acetylhomocysteine to spacer
Spacer terminating in groups which react with thiol groups	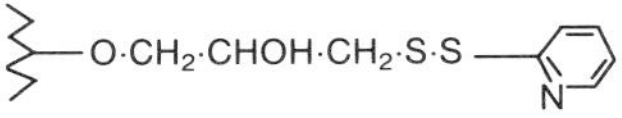 $\sim O \cdot CH_2 \cdot CHOH \cdot CH_2 \cdot S \cdot S$ — 2-pyridyl	R-SH	—S—S—R	Couples reversibly with thiol compounds, including proteins containing thiol groups, by disulfide exchange
	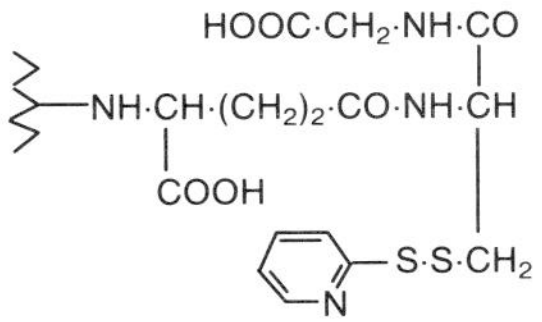 $\sim NH \cdot CH(COOH) \cdot (CH_2)_2 \cdot CO \cdot NH \cdot CH(CO \cdot NH \cdot CH_2 \cdot COOH) \cdot CH_2 \cdot S \cdot S$ — 2-pyridyl	R-SH	—S—S—R	Spacer group is glutathione. Couples reversibly with thiol compounds, including proteins, by disulfide exchange

Continued

Table 2. Selection of agarose matrices with spacers attached, *continued*

Type	Structure	Reactive group of ligand	Mode of coupling of ligand	Additional information
Multivalent spacers: poly-L-lysine	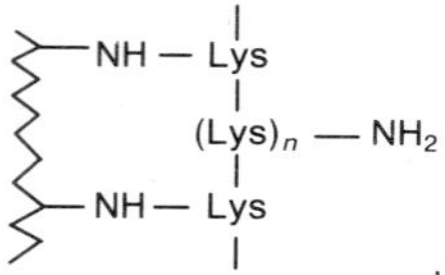 ~NH — Lys —(Lys)$_n$ — NH_2 — Lys — NH~	R-COOH	—NH—CO—R	Poly-L-lysine attached at multiple sites to agarose through ε-amino groups. Multipoint attachment reduces ligand leakage
Polyacrylhydrazido-agarose	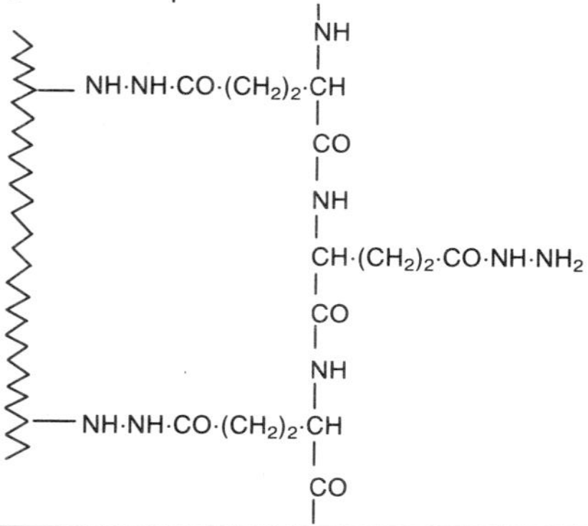 ~NH·NH·CO·$(CH_2)_2$·CH(—NH—)—CO—NH—CH·$(CH_2)_2$·CO·NH·NH_2—CO—NH—CH(·$(CH_2)_2$·CO·NH·NH—~)—CO—			Multipoint attachment limits ligand leakage

Adapted from Dawson, R.M.C. *et al.* (1991) *Data for Biochemical Research.* Oxford University Press, Oxford.

Table 3. Coating and functionalization methods for silica matrices

Matrix	Reagents	Ref.
Aminopropyl-silica	γ-Glycidoxypropyltrimethoxy-silane, 0.1 M sodium acetate buffer pH 5.5, 1–10 mM HCl, acetone	2
Aminopropyl-silica	γ-Glycidoxypropyltrimethoxy-silane, triethylamine, toluene, acetone	4
Diol-silica	γ-Glycidoxypropyltrimethoxy-silane, water, 1–10 mM HCl, acetone	5
Epoxy-silica	γ-Glycidoxypropyltrimethoxy-silane, 0.1 M sodium acetate buffer pH 5.5, 1–10 mM HCl, acetone	2
Epoxy-silica	γ-Glycidoxypropyltrimethoxy-silane, triethylamine, toluene, acetone	4
Isothiocyanatopropyl-silica	3-Isothiocyanatopropyltriethoxy-silane, dichloromethane, 2,6-dimethylpyridine, methanol, diethyl ether, argon	6

Table 4. Activation procedures of diol-silica [2]

Matrix	Reagents
Aldehdye-silica	90% (v/v) acetic acid, sodium periodate, methanol, diethyl ether
Imidazol carbamate-silica	1,1′-Carbonyldiimidazole (CDI), acetonitrile, acetone
Tresyl-silica	Tresyl chloride, acetone, pyridine, 5 mM HCl

Table 5. Ligands which have been immobilized and used for affinity chromatography

Ligands	Separation of
p-Aminobenzamidine	Trypsin, bovine thrombin, urokinase, plasmin, plasminogen activator
Antigens	Specific monoclonal and polyclonal antibodies
Arginine/lysine	Plasmin, plasminogen activators
Avidin	Biotin-containing enzymes
Biotin	Biotin-binding proteins, avidin and avidin derivatives, streptavidin
Boronate (*m*-amino phenyl boronic acid)	Glycoprotein, transfer RNA, nucleotides, nucleosides, catecholamines
Calmodulin	Calmodulin-binding enzymes
Cibracron blue F3G-A	Albumin, NAD-dependent enzymes, cell growth factors, interferons, transferases, polymerases, ribonuclease A
Concanavalin A	Carbohydrates, glycoenzymes, interferon, α-antitrypsin, human alkaline phosphatase, horseradish peroxidase
Dextran blue	Certain kinases, dehydrogenases
Fatty acids	Fatty acid-binding proteins, albumin
Gelatin	Fibronectin
Heparin	Coagulation factors, lipoproteins, lipoprotein lipases, connective tissue proteases, DNA polymerases
Lectins	Glycoproteins
Monoclonal antibodies	Antigens, protein A fusions
Nucleotides	Nucleic acid-binding proteins, nucleotide-requiring enzymes
Phosphoric acid	Phosphatases
Protease inhibitors	Proteases
Protein A/protein G	Immunoglobulins from various species
Steroids	Steroid receptors, steroid binding proteins
Sugars	Lectins, glycosidase
Triazine dyes	Dehydrogenases, kinases, polymerases, interferons, restriction enzymes

Table 6. Lectins commonly used for glycoprotein analysis

Monosaccharide	Lectin (agglutinins)	Source of lectin
N-Acetyl-β-D-glucosamine	WGA	*Triticum vulgare* (wheat germ)
α-*N*-Acetylneuraminic acid	LPA	*Limulus polyphemus* (limulin)
α-L-Fucose	AAA	*Aleuria aurantia*
	LTA	*Lotus tetragonolobus*
	UEA_I	*Ulex europeus I*
β-D-Galactose, *N*-acetyl-β-D-galactosamine	PNA	*Arachis hypogaea* (peanut)
	RCA_I, RCA_{II}	*Ricinus communis*
	SBA	*Glycine max* (soybean)
α-D-Galactose, *N*-acetyl-α-D-galactosamine	DBA	*Dolichos biflorus*
	GSA_I	*Griffonia simplicifolia I*
α-D-Mannose, α-D-glucose	ConA	*Canavalia ensiformis*
	LCA	*Lens culinaris*

Table 7. Determination of ligand concentration

Method	Additional information
Acid or enzymatic hydrolysis	Vigorous treatment of immobilized ligands will hydrolyze the matrix–ligand bond and liberate either free ligand or a degradation product derived from it which may be assayed. Complete hydrolysis of agarose-immobilized ligands can be achieved by heating for 1 h at 100°C in 0.5 M HCl
Difference analysis	The amount of ligand coupled to the gel is estimated by the difference between the total amount of ligand added to the coupling mixture and that recovered after extensive washing. This method is highly inaccurate, especially where only a small proportion of the ligand is covalently attached or when the ligand is sparingly soluble and requires large volumes of wash to remove unbound material. Nevertheless, the method has been used in innumerable cases to give some idea of the concentration of immobilized ligand
Direct spectroscopy	For ligands which adsorb at wavelengths above 260 nm it is possible to estimate the concentration of covalently bound ligand by direct spectroscopy of the gel itself. The gel is suspended in optically clear polyacrylamide, ethylene glycol or glycerol and read against underivatized gel in a double-beam spectrophotometer
Elemental analysis	In some cases elemental analysis can give unambiguous estimates of the ligand concentration. Phosphate analysis has been used for immobilized nucleotides and nucleic acids, while sulfur analysis has been used to estimate sulfanilamide coupled to cyanogen bromide-activated agarose
Radioactive methods	In many cases, radiolabeled ligands have been incorporated into the coupling procedure which in turn have enabled determination of immobilized ligand concentration by difference analysis, hydrolysis or by direct measurements on the gel
Solubilization of gels	Several methods are available for solubilizing derivatized agarose gels which permit quantitative spectrophotometry of the immobilized ligand. Underivatized agarose at concentrations up to those of the moist packed gel can be dissolved by heating to 60°C or above. Derivatized agarose gels may be solubilized by heating to 75°C with either 0.1 M HCl, 0.1 M NaOH–0.1% (w/v) $NaBH_4$, or 50% (v/v) acetic acid. Optimal solubilization conditions are found by trial and error

Table 8. List of elution buffers used for size exclusion HPLC of hydrophobic proteins

Protein(s)	Elution buffer	Ref(s)
Blood platelet membrane proteins	0.5% Berol 185 in 0.1 M triethanolamine–HCl pH 7.4	7
Escherichia coli cytochromes	0.05% Sarkosyl, 0.6 M NaCl, 0.01 M Tris–HCl pH 7.5	8
EIAV, PDGF fragments	0.02 M sodium phosphate pH 6.5 with 6 M guanidinium–HCl 0.01 M sodium phosphate pH 7.0 with 6 M guanidinium–HCl	9, 10
Human erythrocyte ghosts	0.21% SDS, 0.25% CHAPS, 0.08% reduced Triton X-100 or 0.003% Tween-20 in 0.1 M sodium phosphate pH 6.5, 0.1 M NaCl	11
Ia antigens	0.2% Triton X-100, 0.1% triethylamine pH 3.0	12
Immunoprecipitates, tick-borne encephalitis virus, Sendai virus, EIAV, PDGF receptor	0.1% SDS in 0.05 M sodium phosphate pH 6.5 0.1% SDS in 0.1 M sodium phosphate pH 7.0	9, 10, 13–15
Influenza virus	0.1% SDS, or Brij 35 in 0.1 M sodium phosphate pH 7.1	16
Membrane glycoprotein antigen	0.25% sodium deoxycholate in 0.01 M sodium phosphate pH 7.4, 0.15 M NaCl	17
Plasma membrane proteins	0.1% SDS in 0.16 M NaCl with 8 mM sodium phosphate pH 7.0 0.05% CHAPS in 0.01 M Tris–HCl pH 7.1, 0.16 M HCl	18, 19
Sendai virus	0.1% decylpolyethyleneglycol-300 in 0.05 M sodium phosphate pH 6.5 45% acetonitrile in 0.1% HCl	20, 21

CHAPS, 3-(3-cholamidopropyl)dimethylammonio-1-propanesulfonate; EIAV, equine infectious anemia virus; PDGF, platelet-derived growth factor; SDS, sodium dodecyl sulfate.

Table 9. Elution conditions employed for proteins separated on hydroxylapatite columns

Proteins	Eluant
Acidic	Phosphate (30–120 mM) buffer pH 6.8
Neutral	Phosphate (30–120 mM) buffer pH 6.8
Basic, e.g. lysozyme	Phosphate buffer pH 6.8 (phosphate concentration above 120 mM)
Very basic, e.g. lysine-rich histones	Phosphate (500 mM) buffer pH 6.8

Table 10. Conditions required for counter-ion conversion of commercially available ion-exchange columns

Counter-ion		
Original	Required	Method
H^+	Na^+	2 vols 0.1–1.0 M NaOH or 2 vols 3 M NaCl
OH^+	Cl^-	2 vols 0.1–1.0 M HCl or 2 vols 3 M NaCl
Na^+	H^+	30 vols 0.1–1.0 M HCl or 30 vols 3 M NaCl
Cl^-	OH^-	30 vols 0.1–1.0 mM NaOH or 30 vols 3 M NaCl

Table 11. Buffer substances and salts used in ion-exchange HPLC of peptides and proteins

Anion exchange			Cation exchange			
pH range	Substance	Concn (mM)	pH range	Substance	Concn (mM)	Neutral salts
5.0–6.0	Piperazine	20	1.5–2.5	Maleic acid	20	NaCl
7.3–7.7	Triethanolamine	20	2.63–3.63	Citric acid	20	Na_2SO_4
7.6–8.0	Tris	20	3.6–4.3	Lactic acid	50	$(NH_4)_2SO_4$
8.4–8.8	Diethanolamine	20 at pH 8.4 50 at pH 8.8	4.8–5.2	Acetic acid	50	
9.0–9.5	Ethanolamine	20	6.7–7.6	Phosphate	50	
10.6–11.6	Piperadine	20	7.6–8.2	Hepes	50	
11.8–12.0	Phosphate	20	8.2–8.7	Bicine	50	

Table 12. Volatile buffer systems used in ion-exchange chromatography

pH	Substance	Counter-ion
2.0	Formic acid	H^+
2.3–3.5	Pyridine/formic acid	$HCOO^-$
3.0–6.0	Pyridine/acetic acid	CH_3OO^-
6.8–8.8	Trimethylamine/HCl	Cl^-
7.0–12.0	Trimethylamine/CO_2	CO_3^-
7.9	Ammonium bicarbonate	HCO_3^-
8.0–9.5	Ammonium carbonate/ammonia	CO_3^-
8.5–10.0	Ammonia/acid	CH_3COO^-
8.5–10.5	Ethanolamine/HCl	Cl^-

Table 13. Eluotropic series for anions and cations in ion-exchange chromatography[a]

Anions	Cations
Hydroxide (OH^-)[b]	Hydronium (H_3O^+)[b]
Citrate	Trivalent anions
Perchlorate	Barium
Sulfate	Strontium
Oxalate	Calcium
Iodide	Copper (II)
Nitrate	Manganese
Chromate	Silver
Bromide	Cesium
Thiocyanate	Rubidium
Chloride	Potassium
Formiate	Ammonium
Acetate	Natrium
Hydroxide (OH^-)[c]	Hydronium (H_3O^+)[c]
Fluoride	Lithium

[a]Ions are listed in order of decreasing elution strength.
[b]On strong cation and anion exchangers.
[c]On weak exchangers.

Table 14. Additives used with mobile phase for ion-exchange HPLC of peptides and proteins

Additive	Function
Dithioerythreitol	Reducing agent
Ethylenediamine tetraacetic acid (EDTA)	Sequesters calcium and magnesium
Glucose, sucrose, polyethylene glycol	Structural stabilizer
Guanidine hydrochloride	Denaturant, solubilizer
Mercaptoethanol	Reducing agent
Neutral detergents	Solubilizer
Sodium dodecyl sulfate (SDS)	Denaturant, solubilizer
Urea	Denaturant, solubilizer

Table 15. Conditions employed for ion-exchange HPLC of monosaccharides

Analysis	Column type[a]	Functional group	Separation mechanism	Mobile phase
Alditols, monosaccharides	Anion exchange	Quaternary ammonium	Anion exchange	NaOH
Alditols, monosaccharides, oligosaccharides	Cation exchange	Sulfonate, Ca^{2+} form	Ion-moderated partition	Water
Glycoprotein-derived carbohydrates, uronic acids, lactones	Cation exchange	Sulfonate, H^{+} form	Ion-moderated partition	Acetonitrile/water
Hexosamines	Cation exchange	Sulfonate, H^{+} form	Cation exchange	Citrate buffers
Monosaccharides, oligosaccharides	Anion exchange	Amino-propylsilane bonded silica, OH^{-} form	H-bonding between hydroxyls and amines	Acetonitrile/water
Monosaccharides, oligosaccharides	Cation exchange	Sulfonate, Ag^{+} form	Ion-moderated partition	Water
Hexoses, pentoses	Cation exchange	Sulfonate, Pb^{2+} form	Ion-moderated partition	Water
Sialic acids, uronic acids	Anion exchange	Quaternary ammonium		Acetate buffers

[a]Column matrix is polymerized styrene-divinylbenzene.

Table 16. Anion-exchange HPLC of oligonucleotides

Oligonucleotides purified	Solvents
Phosphorylated eicosadecamer	Linear gradient from 0.3 M to 0.6 M NaCl in 30 mM sodium phosphate pH 6.0, 5 M urea
Phosphorylated octadeoxynucleotides	Linear gradient from 1 mM potassium phosphate pH 6.3, 60% (v/v) formamide to 0.18 M potassium phosphate pH 6.3, 60% (v/v) formamide
Restriction fragments	Linear gradient from 0.24 M to 0.66 M NaCl in 0.03 M sodium phosphate pH 6.0, 5 M urea
Single-stranded plasmid DNA	Linear gradient from 0.66 M to 0.84 M NaCl in 0.03 M sodium phosphate pH 6.0, 5 M urea
Supercoiled and linearized plasmid DNA	Linear gradient from 0.66 M to 1.2 M NaCl in 0.03 M sodium phosphate pH 6.0, 5 M urea

Table 17. Organic solvents commonly used in reversed-phase HPLC

Organic solvent	Boiling point (°C)	Freezing point (°C)	Viscosity	UV[a] cut-off (nm)	Additional information[b]
Acetonitrile	82	−42	Low viscosity	190	Aqueous mixtures freeze in all proportions and can be lyophilized; more denaturing than propanols; dried residue can be toxic to cells; good UV transparency (far UV grade essential); expensive
Butan-1-ol	–	–	–	210	Used as part of mixture
Ethanol	78.5	−115	Viscous	210	Less effective eluant than acetonitrile; cheap; not popular
Methanol	64.7	−98.8	Aqueous mixtures are quite viscous	205	Half to two-thirds eluting power of acetonitrile; cheap; rarely used
Propan-1-ol	97.8	−127	Very viscous	210	Good eluting power, greater than acetonitrile; adequate UV transparency
Propan-2-ol	82.3	−89	Very viscous	210	Aqueous mixtures up to 25% freeze; more powerful eluant than acetonitrile, but less powerful denaturant; adequate UV transparency

[a]UV, ultraviolet. 10% transmission [22].
[b]Freezing behavior in dry ice/alcohol freezing mixtures.

Table 18. List of minor components of the mobile phase for reversed-phase chromatography of proteins

Solvent system	Concentration	Additional information
Volatile, low pH		
Heptafluorobutyric acid (HFBA)	10 mM	Miscible in all proportions. Good UV transparency. Slightly longer retention times than TFA
Pyridinium formate (acetate)	0.25–0.5 M pH 4.0–5.0	Used with post-column reaction systems. UV opaque
Triethylammonium formate/acetate	25 mM–0.25 M TEA pH 5.0–6.0	Not as effective as TEAP. Poor UV at high concentration
Trifluoroacetic acid (TFA)	5–50 mM	Miscible in all proportions. Good UV transparency
Nonvolatile, low pH		
0.155 M NaCl/HCl	pH 2.1	Low buffering capacity (up to 80% of CH_3CN). Cl^- corrosive to stainless steel
Phosphoric acid	10 mM	Lower retention times than TFA. Modest buffering power. Add neutral salts to control silanol-related trailing (up to 60% of CH_3CN). Excellent UV transparency
Triethylammonium phosphate (TEAP)	0.1–0.25 M H_3PO_4 pH 2.5–3.0	Does not necessarily give better separations than TFA. Slow to re-equilibrate. Powerful buffering capacity (70–80% of CH_3CN). Neutralizes silanol. Good UV transparency
Neutral		
Ammonium acetate pH 7.0	0.01–0.1 M	Gradient to 15 mM TFA pH 2.0/CH_3CN. Good recovery of intact glycoprotein hormones
Phosphate pH 6.0–7.0	0.01–0.1 M	Limited solubility – maximum 45–50% CH_3CN. Dilute phosphate can be combined with neutral salts. Excellent UV transparency
TEAP pH 7.0	25 mM TEA	Up to 80% CH_3CN. Good UV transparency

Continued

Table 18. List of minor components of the mobile phase for reversed-phase chromatography of proteins, *continued*

Solvent system	Concentration	Additional information
Neutral salts[a]		
Na_2SO_4	60%	Transparent at 210 nm
$(NH_4)_2SO_4$	60%	Transparent at 210 nm
NaCl	80%	Transparent at 210 nm
$NaClO_4$	70%	Transparent at 210 nm

[a] Percentages given are maximum % of CH_3CN.

Table 19. Conditions employed for normal and reversed-phase HPLC of monosaccharides

Analysis	Column type	Separation mechanism	Mobile phase
Monosaccharides, oligosaccharides	Silica, normal phase	H-bonding between hydroxyls and amines	Acetonitrile/water + diaminoalkanes
Pre-derivatized carbohydrates	Silica, normal phase	Polar interactions	Acetonitrile/water
Pre-derivatized carbohydrates, oligosaccharides	Silica, reversed-phase	Hydrophobic interactions	Acetonitrile/water

Table 20. Solvent systems employed for reversed-phase chromatography of lipids and steroids

Molecule	Solvent system	Refs
Free fatty acids	Acetonitrile:tetrahydrofuran:10% phosphoric acid pH 2.0 (50.4:21.6:28)	
	Acetonitrile:tetrahydrofuran:water (45:20:35)	
Fatty acid methyl esters	Acetonitrile:water (95:5)	
	Acetone:acetonitrile:tetrahydrofuran (50:42.4:7.6)	
Fatty acid phenacyl esters and BMC esters	Acetonitrile (or methanol):water gradient from 70–80% to 95–98% acetonitrile	
Triglycerides	Acetone:acetonitrile (63.6:36.4)	23–27
	Acetonitrile:dichloromethane:tetrahydrofuran (60:20:20)	
	Acetonitrile:tetrahydrofuran (50:50)	
	Gradient of 30–90% propionitrile in acetonitrile	
Phospholipids	90–98% methanol or acetonitrile:methanol (70:30–60:40)	
	10–2% potassium phosphate pH 7.4 – 10 M either isocratic elution or methanol gradient from 90–98%	
Bile acids	2-Propanol:8.8 mM potassium phosphate pH 2.5 (32:68)	28–32
	Methanol:water:acetic acid or NaOH pH 4.7 (65:35:3.27)	
	Acetonitrile:methanol:0.03 M potassium phosphate pH 3.4 (10:60:30)	

BMC, 4-bromomethyl-7-methoxycoumarin.

Table 21. Solvent systems used for reversed-phase HPLC of oligonucleotides

Oligonucleotides purified	Solvents
Oligodeoxynucleotides	0.1 M triethylammonium acetate pH 7.0, 1% (v/v) acetonitrile
	0.1 M triethylammonium acetate pH 7.0, 50% (v/v) acetonitrile
Phosphorylated eicosadecamer	Linear gradient from 15% to 30% (v/v) triethylammonium acetate pH 7.0

Table 22. Specific elution procedures used for affinity chromatography

Protein	Adsorbent	Eluant
L-Asparaginase	D-Asparagine–Sepharose	0.001 M D-asparagine
Carbonic anhydrase B	Sulfanilamide–Sephadex	Gradient (0–0.0001 M) of acetazolamide
Lactate dehydrogenase	ε-Aminohexanoyl–NAD^+–Sepharose	Gradient (0–0.05 M) of NAD^+
Ribonucleotide reductase	dATP–Sepharose	Gradients of dATP (0–0.001 M), ATP (0–0.01 M), dAMP (0–0.02 M)
Soybean agglutinin	Sepharose–*N*-ε-aminocaproyl-β-D-galactopyranosylamine	0.5% D-galactose
Threonine deaminase	Isoleucine–Sepharose	0.001 M isoleucine
Thrombin	*p*-Chlorobenzylamido-ε-amino-hexanoyl–Sepharose	0.2 M benzamidine
Trypsin	*p*-Aminophenyl-guanidine–Sepharose	0.001 M benzamidine

Table 23. Some eluants for immunoadsorbents

Hapten	Eluant
Anti-DNP-amino acids	20% (v/v) formic acid
	Glycine–HCl pH 2.8
DNP-lysine	DNP
	DNP-lysine
Galactosidase	8 M urea
Human complement C1	0.2 M 1,4-diaminobutane
IgG	Distilled water
	Sodium thiocyanate
	2.5 M NaI pH 7.5
	2.8 M $MgCl_2$
Insulin	0.01–1 M HCl
	1 M acetic acid pH 2.0
β-Lipoprotein	4 M urea
Serum albumin	1% NaCl pH 2.0
	0.2 M glycine–HCl pH 2.2

DNP, dinitrophenol.

Table 24. Elution conditions for immunopurification

Elution conditions	Ref(s)
Extremes of pH	
0.1 M glycine-HCl pH 2.5	33
1.0 M propionic acid	34
0.15 M NH_4OH pH 10.5	35
Chaotropic salts	
4.0 M $MgCl_2$ pH 7.0	36
2.5 M NaI pH 7.5	37
3.0 M NaSCN pH 7.4	38
Organic chaotropes	
8 M urea pH 7.0	39
6 M guanidine–HCl	40
Organic solvents	
50% (v/v) ethanediol pH 11.5	41, 42
Dioxane/acetic acid	41, 42

Table 25. Equilibration and elution buffers used for fractionating glycopeptides or oligosaccharides

Immobilized lectin	Equilibration buffer	Elution buffer
AAA	PBS	0.05 M fucose in PBS
DSA	PBS	8 mg ml^{-1} of a mixture of *N,N′* diacetylchitobiose and *N,N′,N″*-triacetylchitotriose in PBS
E_4-PHA[a]	PBS	PBS
L_4-PHA[b]	PBS	PBS
RCA_I	PBS	0.15 M Gal in PBS
WGA	PBS	0.1 M GlcNAc in PBS

[a] *Phaseolus vulgaris* erythroagglutinating lectin. [b] *Phaseolus vulgaris* leukoagglutinating lectin. AAA, *Aleuria aurantia* agglutinin; DSA, *Datura stramonium* agglutinin; PHA, phytohemagglutinin; RCA, castor bean agglutinin; WGA, wheatgerm agglutinin; PBS, phosphate-buffered saline; Gal, D-galactose; GlcNAc, *N*-acetyl-D-glucosamine.

Table 26. Buffers for eluting glycoproteins from lectin columns

Lectin	Eluants
ConA	0.1–0.5 M α-methyl mannoside
	0.1-0.5 M α-methyl glucoside
	sugar + 50% ethylene glycol

LCA	0.1–0.5 M α-methyl mannoside
	0.1–0.5 M α-methyl glucoside
	sugar + 50% ethylene glycol
Pea lectin	0.1–0.5 M α-methyl mannoside
	sugar + 50% ethylene glycol
PHA	0.1–0.5 M *N*-acetyl galactosamine
	sugar + 50% ethylene glycol
PNA	0.1–0.5 M lactose
	0.1–0.5 M galactose
	sugar + 50 % ethylene glycol
RCA_{60}	0.1–0.5 M *N*-acetyl galactosamine
	sugar + 50% ethylene glycol
RCA_{120}	0.1–0.5 M lactose
	0.1–0.5 M galactose
	sugar + 50 % ethylene glycol
SBA	0.1–0.5 M *N*-acetyl galactosamine
	sugar + 50% ethylene glycol
	0.1–0.5 M galactose
WGA	0.1–0.5 M *N*-acetyl glucosamine
	N,N′,N″-triacetyl chitotriose
	sugar + 50% ethylene glycol

ConA, concanavalin A; LCA, lentil lectin; PNA, peanut agglutinin; SBA, soybean lectin; and see footnote to *Table 25*.

Table 27. Purification of lectins

Lectins	Chromatography adsorbent	Eluant	Ref.
RCA	Agarose A 0.5	0.2 M galactose in 5 mM phosphate buffer pH 7.2 + 2 M NaCl	43
	Derivatized lactose Bio-Gel P-150	0.2 M lactose in PBS	44
ConA	Sephadex G-100	20 mM glycine–HCl pH 2.0	45
	Derivatized maltose Bio-Gel P-150	0.1 M α-methyl mannoside in PBS	44
LCA	Sephadex G-100	0.1 M glucose	46
	Derivatized maltose Bio-Gel P-150	0.1 M α-methyl mannoside in PBS	44
PNA	Derivatized lactose Bio-Gel P-150	0.1 M lactose in PBS	44
Pea lectin	Sephadex G-150	10 mM glycine–HCl pH 2.0	47
SBA	D-Galactosamine–Sepharose	2.5% lactose	48
WGA	Ovomucoid Sepharose	0.1 M acetic acid	49
	Derivatized di-*N*-acetylchitobiose Bio-Gel P-160	0.1 M *N*-acetylglucosamine in PBS	44

See footnotes to *Tables 25* and *26* for abbreviations.

Table 28. Equilibration and elution buffers used for affinity chromatography

Column type	Equilibration buffer	Elution buffer
NAD^+–Sephadex	10 mM KH_2PO_4–KOH buffer pH 7.5	0–0.5 M KCl gradient in 10 mM KH_2PO_4–KOH buffer pH 7.5
AMP–Sepharose	10 mM tricine–KOH pH 7.5 containing 10 mM glycerol, 5 mM $MgCl_2$, 1 mM EDTA	5 mM NADH in 10 mM tricine–KOH pH 7.5
NAD^+–Sephadex	0.05 M Tris–HCl buffer pH 8.0	0–0.5 M KCl gradient in 0.05 M Tris-HCl buffer pH 8.0
UDP–Sepharose	0.025 M sodium cacodylate buffer, pH 7.4 containing 0.025 M $MnCl_2$, 0.01 M 2-mercaptoethanol	0.025 M sodium cacodylate pH 7.4 containing 0.025 M EDTA, 0.01 M 2-mercaptoethanol, 0.005 M *N*-acetylglucosamine

EDTA, ethylenediamine tetraacetic acid.

Table 29. Chiral barriers incorporated in crown compounds for chiral chromatography

Chiral barrier	Ref(s)
Aromatic bicyclo [3.3.1] nonan derivatives	50, 51
Binaphtyl group	52–55
Biphenanthryl group	56–58
Carbohydrate moiety	59
Chiral carbon atom with a bulky group directly incorporated in crown ring	60, 61
Hericene derivatives	62
Hexahydrochrysese or tetrahydroindenoinden	63
9,9′-spiro-bifluorene	64
Tartaric acid derivatives	65

Table 30. A selection of chiral mobile phases [66]

Chiral mobile phase
0.35 mM acetic acid in dichloromethane
0.35 mM chiral amine:0.35 mM acetic acid in dichloromethane-1-pentanol (99:1, v/v)
0.35 mM dimethyloctylamine:0.35 mM acetic acid in dichloromethane
0.35 mM quinidine:0.35 mM acetic acid in dichloromethane-1-pentanol (99:1, v/v)
0.35 mM quinine ethylcarbonate:acetic acid:dichloromethane-*n*-hexane-1-pentanol (49:50:1, by vol.)
0.35 mM quinine:acetic acid in dichloromethane-1-pentanol (99:1, v/v)
2.2 mM (+)-10-camphorsulfonic acid in dichloromethane-1-pentanol (199:1, v/v)
2.5 mM benzoxycarbonyl-glycyl-proline (ZGP):0.2 mM triethylamine in dichloromethane (80 ppm H_2O)
2.5 mM ZGP:1.0 mM triethylamine in dichloromethane (80 ppm H_2O)
20 mM phosphate buffer pH 7.0:1.33 M 2-propanol
50 mM phosphate buffer pH 8.9
2-propanol:heptane (7.5:92.5, v/v)
Acetonitrile-aqueous solution:0.25 M ammonium acetate:0.1 mM copper sulfate
Chiral salts in dichloromethane-1-pentanol
Ethanol
Ethanol:water[a]
Hexane:2-propanol:acetonitrile (850:105:75, by vol.)
Hexane:2-propanol[a]
Isopropanol:hexane (10:90, v/v)
MeCN
MeCN:buffer[a]
MeCN:water[a]
Methanol
Methanol:water[a]
Methanol:water[a] containing crown ethers

[a]Different ratios are used.

Table 31. Selection of chiral additives

Applied enantiomers	Chiral additive	Ref(s)
R—CHCOOH(NH₂) Free amino acids	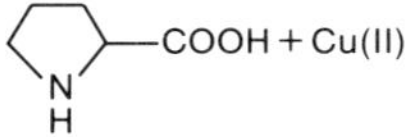 (pyrrolidine ring, NH)—COOH + Cu(II); L or D-proline	67, 68 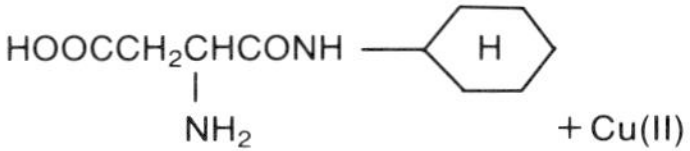
	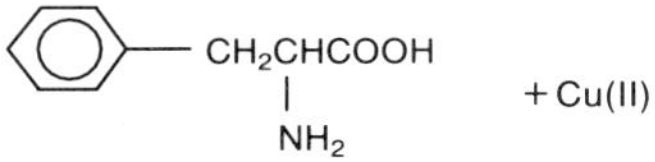 C_6H_5—$CH_2CHCOOH$ (NH_2) + Cu(II); L-Phenylalanine	69 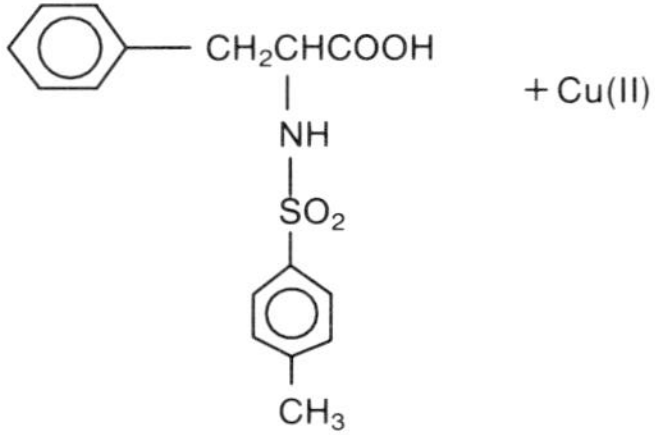
	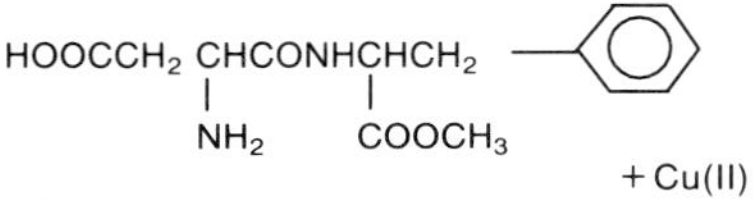 $HOOCCH_2$ $CHCONHCHCH_2$—C_6H_5 (NH_2, $COOCH_3$) + Cu(II); L-Aspartyl-L-phenylalanine methyl ester	70
	$HOOCCH_2CHCONH$—(cyclohexyl, H) (NH_2) + Cu(II); L-Aspartyl cyclohexyl amide	71–73
	C_6H_5—$CH_2CHCOOH$ (NH—SO_2—C_6H_4—CH_3) + Cu(II); N-(p-toluenesulfonyl) L-phenylalanine	74, 75

Continued

Table 31. Selection of chiral additives, *continued*

Applied enantiomers	Chiral additive	Ref(s)
	CH_3–C_6H_4–SO_2–NH–CH(C_6H_5)COOH + Cu(II) *N*-(*p*-toluenesulfonyl) D-phenylglycine	76, 77
	R-CHCOOH, N(R′)(R′) + Cu(II) *N*,*N*-Dialkyl-L-amino acids	78

Applied enantiomers	Chiral additive	Ref(s)
	C_6H_5–$CH_2CHCOOH$, N(R)(R′) + Cu(II) *N*-Methyl-L-phenylalanine ($R = CH_3$, $R' = H$) *N*,*N*-Dimethyl-L-phenylalanine ($R = R' = CH_3$)	83
R-CHCOOH–NH–SO_2–naphthyl–$N(CH_3)_2$ Dns-amino acids	R–CH(NH_2)–CH_2–N($(CH_2)_7CH_3$)–CH_2CH_2–NH_2 + Zn (II), Cd (II), Ni (II), Cu (II), Hg (II) $R = C_2H_5$, $CH(CH_3)_2$, $CH_2CH(CH_3)_2$ L-2-Alkyl-4-*n*-octyl-diethylene triamine	84,85

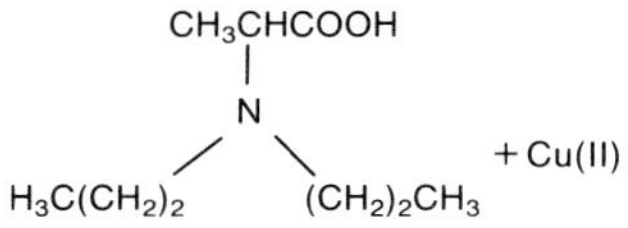

$CH_3CHCOOH$ — N — $H_3C(CH_2)_2$, $(CH_2)_2CH_3$ + Cu(II) — 79–81

N,N-di-*n*-propyl
L-alanine

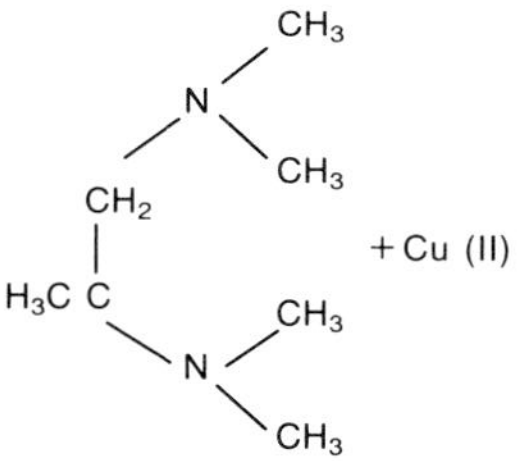

$(CH_3)_2N$–CH_2–$C(H_3C)$–$N(CH_3)_2$ + Cu (II) — 82

N,N,N′,n′-Tetramethyl-
(R)-propane, 2-diamine

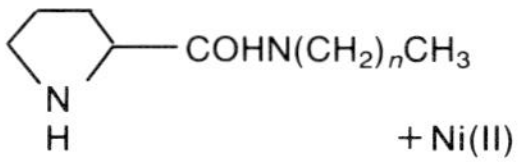

(pyrrolidine ring, N–H)–$COHN(CH_2)_nCH_3$ + Ni(II) — 85, 86

L-Prolyl-*n*-octylamide
($n=7$)
L-Prolyl-*n*-dodecylamide
($n=11$)

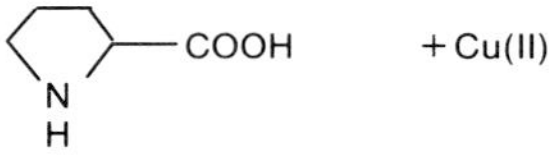

(pyrrolidine ring, N–H)–COOH + Cu(II) — 87–89

L-Proline

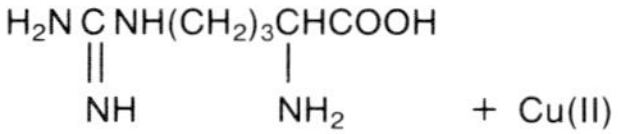

$H_2NC(=NH)NH(CH_2)_3CH(NH_2)COOH$ + Cu(II) — 88

L-Arginine

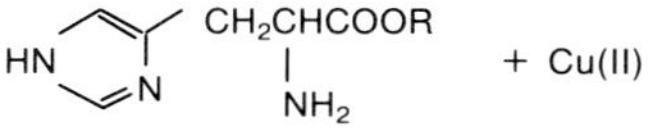

(imidazole ring, HN, N)–$CH_2CH(NH_2)COOR$ + Cu(II) — 88, 90–92

L-Histidine ($R=H$)
L-Histidine methyl ester
($R=CH_3$)

Continued

Table 31. Selection of chiral additives, *continued*

Applied enantiomers	Chiral additive	Ref(s)	Applied enantiomers	Chiral additive	Ref(s)
R—CH(OH)COOH Hydroxy acids	C_6H_5—$CH_2CH(NRR')COOH$ + Cu(II) L-Phenylalanine (R = R′ = H)	93, 94	R—CH(OH)—CH_2NR′R″ β-Amino alcohols	HO(CONH$(CH_2)_7CH_3$)CH—CH(OH)COOH + Cu(II) (*R*,*R*)-Tartaric acid mono-*n*-octylamide	95
	N-Methyl-L-phenylalanine (R = CH_3, R′ = H) *N*,*N*-Dimethyl-L-phenylalanine (R = R′ = CH_3)	83	9-substituted guanine (H_2N, HN, C=O, N) with $CH_2CH(OH)CH(OH)CH_2$ side chain 9-(3,4-dihydroxy-butyl)guanine	C_6H_5—$CH_2CH(NH_2)COOH$ + Cu(II) L-Phenylalanine	96
	Pyrrolidine-2-COOH (N–H) + Cu(II) L-Proline	94			

Adapted from Nimura, N. (1989) in *Chiral Separations by HPLC; applications to pharmaceutical compounds* (A.M. Krstulovic, ed.). Ellis Horwood Ltd, Chichester.

Table 32. Mobile phases used for measuring amines and their metabolites by electrochemical detection

Amines detected[a]	Mobile phase	Ref(s)
DOPAC, 5HT, tryptophan, 5HIAA, HVA, DOPA, 5HTP	0.1 M acetate/citrate buffer pH 4.6 with 10% (v/v) methanol	97, 98
MHPG, NA, A, DHBA, DOPAC, 5HIAA, DA	0.1 M NaH_2PO_4, 0.1 mM EDTA, 1.0 mM sodium octylsulfate, 9% methanol pH 3.6	99–101
NA, A, DA, DOPAC, 5HT, 5HIAA, HVA	0.15 M NaH_2PO_4, 0.1 mM EDTA, 0.5 mM sodium octanesulfonic acid, 14% methanol pH 3.4	
NA, A, DA, DOPAC, 5HT, 5HIAA, HVA	0.1 M NaH_2PO_4, 1.0 mM EDTA, 1.0 mM sodium octanesulfonic acid, adjust to pH 4.0–4.4 with citric acid. Mix with acetonitrile	102
3-MT	0.1 M trichloroacetic acid, adjust to pH 3.0 with 1 M sodium acetate, 0.1 M EDTA, 20% methanol	100

[a]A, adrenaline; DA, dopamine; DHBA, 3,4-dihydroxybenzylamine; DOPA, 3,4-dihydroxyphenylalanine; DOPAC, 3,4-dihydroxyphenylacetic acid; 5HIAA, 5-hydroxyindoleacetic acid; 5HT, 5-hydroxytryptamine; 5HTP, 5-hydroxytryptophan; HVA, homovanillic acid; MHPG, 3-methoxy-4-hydroxyphenylglycol; 3-MT, 3-methoxytryptamine; NA, noradrenaline. EDTA, ethylenediamine tetraacetic acid.

Table 33. Mobile phases used for liquid chromatography of vitamins

Vitamin	Mobile phase	Ref.
Provitamin A carotenoids	Acetonitrile:dichloromethane:methanol (70:20:10)	103
Retinoic acid	*n*-hexane:acetonitrile:acetic acid (99.5:0.2:0.3)	104
Retinol	100% methanol	
Retinyl esters	100% methanol	105
Riboflavin (Vitamin B_2)	34% (v/v) methanol in water	106
Thiamin (Vitamin B_1)	45% (v/v) methanol in 0.05 M sodium citrate pH 4.0 containing 10 mM sodium 1-octanesulfonate	107
Vitamin C	20% (v/v) methanol in 0.08 M potassium dihydrogen phosphate pH 7.8	108
Vitamin D metabolites	Methanol:water (7:3)	109

Chapter 6 DETECTION AND ANALYSIS

1 Introduction

Another key factor in the successful application of liquid chromatography is the detection and analyses of molecules in the eluate fractions. Consequently, a vast amount of research time and energy has been devoted to the development of the ideal detection method and/or unit. The topics covered in this chapter include a selection of various analytical and detection methods, analyte properties, and the different types of reagents used for derivatization and detection. The detection methods employed in high-pressure liquid chromatography (HPLC) differ little from those used in low-pressure liquid chromatography, with the exception that as a result of the rapid separation process, a faster detector response speed is necessary with HPLC.

2 Detectors

The once laborious and time-consuming sequential analyses of collected eluate fractions have been replaced by an on-line analyte detector module. In most cases, the detector module, through which the eluate from the column flows, generates a continuous electrical output signal that is a function of the concentration or the mass of the analyte in the mobile phase. This electrical signal is passed directly to a chart recorder providing either a record of the separation as a function of time or a chromatogram. The quality of the separation is assessed from the chromatogram. However, the reader should be aware that, in addition to the column, the quality of the separation is dependent on the detector response time, the analyte electrical response signal to detector noise ratio and the flow cell design; the flow cell is that part of the detector module which holds the aliquot of the eluate which is actually being monitored by the detector device. A comparison of the many different types of detector modules that are commercially available is given in *Table 1*; the most popular of the detectors are those involving ultraviolet (UV) photometers. For in-depth material and reviews on detectors, the reader should consult Scott [1] and Ben-Bassat and Grushka [2]. Enzyme peak shifts, and UV and electrochemical properties of various analytes are also listed (*Tables 2–5*).

Furthermore, in the case of size exclusion chromatography, due to the complexity of the original sample it can not be assumed that the separate UV absorbing peaks used to monitor the separation are due to a single analyte. Further analysis of the eluate fractions may need to be performed. A summary of the available techniques used to analyse eluate fractions further is given in *Table 6*.

3 Detection reagents and derivatization methods

As previously stated (Chapter 4, Section 3), the characteristics of many native molecules may not permit direct estimation in solution and so, to aid detection and quantification of particular molecules, it is often necessary to derivatize the sample components to a modified form which then lends itself to detection. The derivatization process must be compatible with the mobile phase used. Quantification from the detector's response is made by direct comparison with standards. A selection of detection reagents and derivatization techniques are listed in *Tables 7–9*. Post-column derivatization should be the method of choice if a greater sensitivity than that accomplished by direct detection is required.

4 Radioactive detection

Molecules which have been radiolabeled can be monitored by scintillation counting of the eluate fractions. The different types of scintillation counting and the counting efficiency of Cerenkov counting are given in *Tables 10* and *11*.

Table 1. Comparison of commercially available detector modules

Detector	Sensitivity	Applicability	Common examples
UV photometric fixed wavelength variable wavelength diode array	High	High sensitivity for strong UV absorbers. Limited to those analytes which absorb in the UV (or visible region) of the electromagnetic spectrum. The range of applications may be considerably increased if analytes are modified by either pre- or post-column chemical derivatization	Proteins and peptides; amino acids; nucleic acids and components; oligonucleotides; oligosaccharides; glycoproteins; proteoglycans; fatty acids and esters; eicosanoids; triglycerides; phospholipids; bile acids; steroids; vitamins [3–5]; porphyrins; bile pigments; organic acids; pesticides; phenols
Fluorescence	High	Limited use as so few analytes naturally fluoresce. The range of applications may be considerably increased if analytes are chemically modified to form fluorophores by either pre- or post-column derivatization	Steroids [6]; vitamins; porphyrins
Refractive index	Moderate	Useful for a wide range of analytes. However, is sensitive to small temperature ($\pm 0.01^{\circ}C$) and pressure changes and so can only be used for isocratic (elution at constant composition) separations	Monosaccharide; oligosaccharides [7–10]; fatty acids and esters; triglycerides; bile acids; steroids [11]
Electrochemical amperometric	High	Only applicable to analytes which can be oxidized or reduced at a suitable working electrode	Small peptides; amino acids; nucleic acid components; oligosaccharides [12–17]; proteoglycans; steroids [18]
Electrochemical conductimetric	Medium	Used most widely for ion-exchange HPLC. Applicable to anions and cations with pK_a or $pK_b < 7$	

Continued

Table 1. Comparison of commercially available detector modules, *continued*

Detector	Sensitivity	Applicability	Common examples
Fluorescence, β-induced	High	Limited range of applicability, e.g. aromatic hydrocarbons	
Refractive index thermal lens, laser-induced	Moderate	Useful for wide range of analytes. Suitable for isocratic separations	
Interferometry	High	Applicable to wide range of analytes. Only suitable for isocratic separations	
Light-scattering		Detectors based on these effects are suitable for certain classes of analytes, e.g. polymers and large molecules or optically active molecules	Bile pigments
Circular dichroism		As for light-scattering	
Optical rotating dispersion		As for light-scattering	
Radioactivity	High	Expensive	Proteins, nucleic acid components; oligosaccharides [19]; proteoglycans; eicosanoids; steroids
Infra-red spectroscopy	Limited	Normally restricted to nonaqueous eluant systems. Primarily used for the identification of specific structural features	Oligosaccharides [20, 21]; triglycerides
Nuclear magnetic resonance	Limited	Expensive. Applicable to a wide range of analytes	Oligosaccharides [22–25]
Electron spin resonance	High	Only applicable to those analytes which can be induced to form free radical derivatives by pre- or post-column derivatization	
Mass spectrometry	Good/high	Expensive. Only applicable to analytes of low molecular mass (10^2–10^3)	Oligosaccharides [26]; triglycerides; steroids

Table 2. Enzyme peak shifts used to identify peaks for nucleic acid components [27]

Peak	Enzyme	Reaction pH	Product
Adenosine	Adenosine deaminase	7.5	Inosine
Uric acid	Uricase	8.5	Allantoin
NMP	5′-Nucleotidase	7.4	Nucleoside
Trytophan	Tryptophanase	8.3	Indole
cAMP	Phosphodiesterase cAMP	7.5	AMP

Table 3. UV properties of nucleotides, nucleosides and bases

				Ratio	
	pH	Max nm[a]	$E \times 10^{-3}$	254:220	254:270
ATP	2	257	14.7	3.05	1.77
GTP	1	256	12.4	3.80	1.81
CTP	2	280	13.0	1.03	0.52
UTP	2	262	10.0	33.1	1.40

[a]Maximum and E are very similar for base, nucleoside, NDP, and NMP.

Table 4. Electrochemical activity; peak oxidation potentials of amino acids and neuropeptides[a] at pH 4.6 [28]

	V^b
Amino acids	
Cysteine	0.90–1.00
Tryptophan	0.88
Tyrosine	0.84
Neuropeptides	
ACTH1-24	0.80–0.88
Leu-enkephalin	0.84
LH-releasing hormone	0.83
Met-enkephalin	0.84
β-MSH	0.80–0.86
Neurotensin	0.78
Oxytocin	0.82
Somatostatin	0.86–0.90
Vasopressin	0.82

[a]Peptides containing neither tryptophan nor tyrosine are not electroactive.
[b]*V* is the potential at which the maximum current was generated.
ACTH, adrenocorticotropic hormone; LH, luteinizing hormone; MSH, melanocyte-stimulating hormone.

Table 5. Electrochemical activity of bases, nucleosides and nucleotides[a] at pH 4, 7 and 9

	pH	Base	Nucleoside	NMP	NDP	NTP
Adenine	4	+	+	–	–	–
	7	++	++	–	–	–
	9	++	++	–	–	–
Cytosine	4	–	–	–	–	–
	7	–	–	–	–	–
	9	–	–	–	–	–
Guanine	4	+++	+	++	++	++
	7	+++	+	++	++	++
	9	+++	+	+	+	+
Hypoxanthine	4	+	–	–	ND	ND
	7	++	–	–	ND	ND
	9	++	–	–	ND	ND
Thymine	4	ND	–	ND	ND	ND
	7	ND	–	ND	ND	ND
	9	ND	–	ND	ND	ND
Uracil	4	+	ND	–	ND	–
	7	+	ND	–	ND	–
	9	+	ND	–	ND	–
Xanthine	4	++	+	++	++	++
	7	++	+	++	++	++
	9	++	+	+	+	+

[a]Determined by flow injection analysis using an amperometric electrochemical detector equipped with a glassy carbon electrode at 1.1 V vs. Ag/AgCl reference electrode.

–, not electroactive; +, weakly electroactive; ++, moderately electroactive; +++, strongly electroactive; ND, not determined.

Table 6. Analytical methods employed after size exclusion chromatography of proteins

Analytical method	Comments
Enzymatic activity	Used to determine function and stability of protein. It may be necessary to perform chromatography of sample in the cold and at the highest possible flow-rate
Immunological activity	Eluate fractions may be used as antigens in immunological assays. The remaining immunological activity after chromatography can be determined using an enzyme-linked immunosorbent assay (ELISA)
SDS–polyacrylamide gel electrophoresis (SDS–PAGE)	Used to estimate molecular weights of the proteins in UV absorbing fractions. In combination with silver staining may also be used to assess purity

Table 7. Comparison of detection reagents used for amino acids

	Dansyl chloride (DNS-Cl)	Fluorescamine	Ninhydrin	*o*-Phthalaldehyde thiol (OPT-thiol)
Solvent	Anhydrous	Anhydrous	Aqueous	Aqueous
Incubation conditions	30 min at 37°C	< 1 sec at room temperature	20 min at 100°C	1–2 min at room temperature
Detection	Fluorescence 340/510	Fluorescence 390/475	Absorbance 550, 470	Fluorescence 340/455
Sensitivity	Good, 10–50 pmol	Good	Moderate to good, > 100 pmol	Very good, < 10 pmol
Secondary amines	Yes[a]	No	Yes[a]	No
Additional information	Cons: reacts with water, fluorescence of DNS-OH[a] side reactions	Pros: very rapid reaction Cons: reacts with water	Pros: does not react with water Cons: NH_3 interference, stability of reagent	Pros: does not react with water, rapid reaction, electrochemical activity Cons: will not react with secondary amines or amino acids unless they have been converted to primary amine derivatives, derivatives are not stable

[a]At different wavelengths.

Table 8. Detection reagents

Functional group	Reagent	Detection	Additional information	Ref(s)
Carboxylic acid	*p*-Bromophenacyl-8$^{TM a}$	UV	Pre-column; nanomole detection levels: λ_{max} = 260 nm	29–35
	p-Nitrobenzyl-8$^{TM a}$	UV	Pre-column; picomole detection levels: λ_{max} = 265 nm	36, 37
	Phenacyl-8$^{TM a}$	UV	Pre-column; nanomole detection levels: λ_{max} = 250 nm	29–35
Primary amine	DABITC	Visible	Pre-column; picomole detection levels: λ_{max} = 436 nm	38–46
	Dabsyl chloride	Visible	Pre-column; nanomole detection levels: λ_{max} = 436 nm	42, 47–52
	Dansyl chloride	Fluorescence, UV	Pre-column; picomole detection levels with fluorescence: λ_{ex} = 360 nm, λ_{em} = 470 nm, λ_{max} = 250 nm	53–59
	FDAA, Marfey's	UV	Pre-column; nanomole detection levels: λ_{max} = 340 nm	60–62
	Fluoraldehyde	Electrochemical, fluorescence	Pre- or post-column; picomole detection levels with both methods of detection, GC + 0.5–1.0 V; λ_{ex} = 360 nm, λ_{em} = 455 nm	63–68
	Ninhydrin	Visible	Post-column; nanomole detection levels: λ_{max} = 570 nm	69
	PITC	UV	Pre-column; picomole detection levels: λ_{max} = 254 nm	70, 71
	TNBSA	Electrochemical, UV	Pre- or post-column; nanomole detection levels with both methods of detection, GC −0.85 V; λ_{max} = 250 nm	72, 73
Secondary amine	Ninhydrin	Visible	Post-column; nanomole detection levels: λ_{max} = 440 nm	69
	PITC	UV	Pre-column; picomole detection levels: λ_{max} = 254 nm	70, 71
Phenol	Dansyl chloride	Fluorescence, UV	Pre-column; picomole detection levels with fluorescence: λ_{ex} = 360 nm, λ_{em} = 470 nm, λ_{max} = 250 nm	53–59

Thiol	SBF-chloride	Fluorescence	Pre-column; picomole detection levels: λ_{ex} = 390 nm, λ_{em} = 510 nm	74

[a]Supplier: Pierce (PRC).

DABITC, 4-*N*,*N*-dimethylaminoazobenzene-4′-isothiocyanate; dabsyl chloride, 4-*N*,*N*-dimethylaminoazobenzene-4′-sulfonyl chloride; dansyl chloride, 5-*N*,*N*-dimethylamino-naphthalene-1-sulfonyl chloride; FDAA, 1-fluoro-2-4-dinitrophenyl-5-L-alanine amide; PITC, phenylisothiocyanate; SBF-chloride, ammonium-4-chloro-7-sulfobenzofurazan; TNBSA, 2,4,6-trinitrobenzene-sulfonic acid.

Table 9. Post-column detection of carbohydrates

	Reagent	Ref(s)
Amino sugars	Ninhydrin	75, 76
	o-Phthalaldehyde	77
Carbohydrates	2,2′-Bicinchoninate	78
	2-Cyanoacetamide	79–81
	Ethanolamine-boric acid	82
Carbohydrates, sugar alcohols	Ammoniacal cupric sulfate	83
	0.5 M NaOH	84
Reduced and reducing sugars	4-Aminobenzoylhydrazide	85
	Tetrazolium blue	75, 86
Sialic acids	300 mM NaOH	87

Table 10. Scintillation counting

Procedure	Scintillant	Additional information
Cerenkov counting		High energy (> 0.5 MeV) β-particles traveling through water cause the polarization of molecules along their trajectory, which then emit photons of light (350–600 nm) as their energy returns to the ground state (Cerenkov effect). Counting is achieved in aqueous buffer in a scintillation counter without added scintillation fluid
Counting of finely dispersed or solvent-soluble substances	(i) Scintillation-grade toluene or xylene containing 5 g l^{-1} PPO and 0.1 g l^{-1} POPOP (ii) Scintillation-grade toluene or xylene containing 10 g l^{-1} PBD or 15 g l^{-1} butyl PBD	Chemical quenching and color quenching can both occur. Particulate material will also tend to reduce the counting efficiency
Counting of aqueous samples	(i) 600 ml toluene or xylene (sulfur-free or scintillation grades), 340 ml surfactant (Triton X-100 or Triton X-114), 5 g PPO, 0.1 g POPOP (ii) 575 ml xylene, 250 ml Triton X-100 (X-114), 140 ml ethanol, 35 ml ethylene glycol, 3 g PPO, 0.2 g POPOP	(i) PPO and POPOP can be replaced by 15 g PBD or 10 g PBD. It should be noted that separation into two phases can occur with a marked reduction in counting efficiency [88] (ii) This system can accept up to 23% of water with no phase separation and gives a linear relationship between counting efficiency and external ratio over wide quenching ranges [89]

Gamma counting	Solid scintillator, usually NaI containing thallium	Scintillator surrounds the sample and light (350–500 nm) is detected by photomultipliers. No preparation of the sample is required and the counting efficiency is not affected by the color of the sample or by chemical quenching

PPO, 2,5-diphenyloxazole; POPOP, 1,4-di-(2-(5-phenyloxazolyl))-benzene; PBD, 2-phenyl-5-(4-biphenylyl)-1,3,4-oxadiazole; butyl PBD, 2-(4′-*t*-butylphenyl)-5-(4″-biphenylyl)-1,3,4-oxadiazole.

Table 11. Counting efficiency using Cerenkov counting

Radioisotope	E_{max} (MeV)	% of β-spectrum above 0.5 MeV	Counting efficiency (%)
^{24}Na	1.39	60	40
^{32}P	1.71	80	50
^{36}Cl	0.71	30	10
^{42}K	3.58	90	80

Chapter 7 TROUBLESHOOTING GUIDE

The following table lists symptoms, causes and appropriate remedies suggested for problems experienced in liquid chromatography.

Table 1. Troubleshooting guide[a]

Problem	Cause	Remedy
1. Bubbles or cracks in bed	a. Eluant not properly de-gassed b. Column either packed or stored at cool temperature and then warmed up c. Air leak in column	a. De-gas eluant thoroughly b. Passing de-gassed buffer upwards through column may remove bubbles. Re-pack column c. Check all connections for leaks. Re-pack column
2. No or reduced flow through the column	a. Air lock in outlet tubing or bottom-piece b. Bed surface blocked by precipitated sample c. Bed compressed d. Microbial growth on column	a. Reconnect carefully making sure that all inlets and outlets are closed and free of air pockets b. Remove contaminated gel from bed surface. Agitate top 1–2 cm and allow to settle. Change pre-column c. Re-pack or replace column d. Columns should be stored at +4–8°C in the presence of 0.05% NaN_3

3. Gel beads in eluant	a. Bed support loose or broken b. Column operated at too high pressure	a. Re-fasten or replace b. Do not exceed the recommended operating pressure for the gel medium
4. Distorted bands as sample runs into and through the bed	a. Air bubble at top of column b. Particles in sample or eluant c. Uneven bed surface or particles on the bed surface d. Column poorly packed	a. See 2a b. Filter or centrifuge sample and eluant c. See 2c d. Check packing by running a colored compound and observe band. Re-pack carefully. Avoid packing at very high pressures
5. Sample components elute in the equilibrium phase	a. Sample not sufficiently hydrophobic to adsorb to column b. Unsuitable pH c. Adsorption of impurities to the column d. Concentration of organic modifier in the initial mobile phase is too high e. Column improperly equilibrated	a. Increase the concentration of ion pairing agent. Change to column with a more hydrophobic immobilized ligand or change to an organic modifier with less efficient elution properties b. Adjust pH so that the sample binds c. Clean and regenerate the column d. Decrease the concentration of organic modifier e. Equilibrate until the baseline is stable
6. Sample component elutes in the washing step	a. Ionic strength of initial buffer is too high b. Ionic strength of sample is too high c. Detergents and additives have adsorbed to the column d. Column improperly equilibrated	a. Decrease ionic strength of initial buffer b. Buffer exchange sample c. Clean column d. See 5e
7. Sample does not elute	a. pH of solutions is incorrect	a. Prepare new solutions

Continued

Table 1. Troubleshooting guide[a], *continued*

Problem	Cause	Remedy
8. Protein does not elute in the salt gradient	a. pH of buffer is incorrect b. Ionic strength is too low	a. Use buffer pH closer to pI of protein b. Use a more concentrated limit buffer
9. Sample does not elute in the organic solvent gradient	a. Unsuitable pH b. Eluting power of organic modifier is too low	a. Adjust pH such that sample is not denatured b. Change to organic modifier with more efficient elution properties. Consider changing to a reversed-phase column with a less hydrophobic immobilized ligand
10. Sample component elutes at unexpected position	a. Ionic interaction between protein and matrix b. Hydrophobic interactions between sample and matrix c. Column is dirty	a. Maintain ionic strength above 0.05 M b. Reduce salt concentration, add suitable detergent or organic solvent c. Clean and regenerate column
11. Elution profile can not be reproduced	a. Sample has altered b. Sample mass or volume is different from previous occasion c. Sample improperly filtered d. Proteins or lipids have precipitated on column e. Incomplete column equilibration f. Incorrect buffer pH and ionic strength	a. Prepare fresh sample b. Keep mass and volumes constant for each run c. Regenerate the column, filter sample and repeat column separation d. Use elution conditions which stabilize sample e. See 5e f. Prepare new solutions
12. Poor resolution of sample component	a. Sample mass or the volume is too large b. Sample is too viscous c. Sample improperly filtered	a. Decrease the sample load or sample volume. Apply sample carefully b. Dilute sample with elution buffer

Problem	Cause	Remedy
	d. Proteins or lipids precipitated on column	c. See 11c
	e. Aggregate formation of proteins in sample and strong binding to gel	d. See 11d
	f. Flow-rate too high or detector cell volume is too big	e. Use urea or zwitterions
	g. Large mixing spaces (dead volumes) in or after column	f. Reduce flow-rate or change the flow cell
	h. Uneven temperature in the bed	g. Reduce dead spaces
	i. Gradient slope is too steep	h. Use column with a water-jacket
	j. Column is dirty or poorly packed	i. Use a shallower gradient
	k. Column is too short	j. See 10c and 4d
	l. Microbial growth in the column	k. Longer columns provide better resolution in analytical separation
		l. See 2d
13. Poor recovery of sample amount in eluted fractions	a. Sample degraded by proteases or nucleases	a. Add protease or nuclease inhibitors to buffers
	b. Sample precipitates	b. Adjust salt concentration
	c. Adsorption	c. Add ethylene glycol (10%) to buffers to inhibit nonspecific adsorption
	d. Hydrophobic proteins	d. Use chaotropic salts for elution
	e. Microbial growth on column	e. See 2d
14. Poor recovery of activity	a. Sample may not be stable in the chosen eluant or solvent and is inactivated	a. Change eluant
	b. Enzyme sample separated from cofactor or similar	b. Test by pooling fractions and repeat assay
	c. Microbial growth on column	c. See 2d
15. Leading or very rounded peaks observed in chromatogram	a. Overloaded column	a. Decrease sample load
	b. Column is poorly packed	b. See 4d
	c. Column needs regeneration	c. Clean and regenerate the column. If necessary, replace column

Continued

Table 1. Troubleshooting guide[a], *continued*

Problem	Cause	Remedy
16. Peaks too small	a. Sensitivity range incorrectly set on detector or recorder b. Sample absorbs poorly at chosen wavelength	a. Adjust range b. Use different wavelength
17. Strange peaks observed	a. Buffer impurities	a. Filter buffer through pre-column
18. Column is clogged	a. Presence of particulates, lipoproteins or protein aggregates b. Precipitation of proteins in the column caused by removal of stabilizing agents during separation c. Column filter is clogged d. Microbial growth in the column	a. Prior to chromatography, precipitate with 10% dextran sulfate or 3% polyvinylpyrrolidone. Clean or replace column b. Adjust the eluant to maintain stability. Clean and regenerate column c. Replace filter. Ensure that all solutions and samples are filtered d. See 2d

[a]Adapted from *Gel Filtration: Principles and Methods*, 6th Edn, 18-1022-18 and *Ion Exchange: Principles and Methods*, AA Edn, 18-1114-21 with permission from Pharmacia Biotech (PMB).

Chapter 8 MANUFACTURERS AND SUPPLIERS

Many of the larger companies have subsidiaries in other countries while most of the smaller companies market their own products or through agents. The name of a local supplier can be obtained by contacting the relevant company listed here. The numbers bracketed are area, freephone or freefax code numbers; international dialing codes have not been listed. UK (0800) and USA (800) freephone or freefax numbers can only be used in the corresponding countries.

ACL **Anachem Ltd**, Anachem House, 20 Charles Street, Luton, Beds LU2 0EB, UK.
Tel (01582) 456666.
Fax (01582) 391768.

BKI **Beckman Instruments (UK) Ltd**, Oakley Court, Kingsmead Business Park, London Road, High Wycombe HP11 1JV, UK.
Tel (01494) 441181.
Fax (01494) 463836.

BRL **Bio-Rad Laboratories Ltd**, Bio-Rad House, Maylands Avenue, Hemel Hempstead, Herts HP2 7TD, UK.
Tel (01442) 232552, (0800) 181134.
Fax (01442) 259118.
Bio-Rad Life Science Group, 2000 Alfred Nobel Drive, Hercules, CA 94547, USA.
Tel (510) 741 1000, (800) 424 6723.
Fax (510) 741 5800, (800) 879 2289.

BTL **BAS Technicol Ltd**, Adcroft Street, Higher Hillgate, Stockport, Cheshire SK1 3HZ, UK.
Tel (0161) 477 7020.
Fax (0161) 480 6090.

DPA **Dyno Particles A.S.**, PO Box 160, N-2001, Lillestrøm, Norway.
Tel 6389 7100.
Fax 6389 7472.

DPL **Du Pont Ltd**, Wedgewood Way, Stevenage, Herts SG1 4QN, UK.
Tel (01438) 734015.
Fax (01438) 734049.

ENI **Schiapparelli Biosystems, Inc. (formerly Electro-Nucleonics, Inc.)**, 368 Passaic Avenue, Fairfield, NJ 07004, USA.
Tel (201) 882 8630.
Fax (201) 227 6700.

JCL **Jones Chromatography Ltd,** New Road, Hengoed, Mid Glamorgan CF8 8AU, UK.
Tel (01443) 816991.
Fax (01443) 816552.

JTB **Mallinckrodt Baker, Inc. (formerly J.T. Baker)**, 222 Red School Lane, Phillipsburgh, NJ 08865, USA.
Tel (908) 859 2151.
Fax (908) 859 9454.

LSI **Life Sciences International (UK) Ltd**, Unit 5, The Ringway Centre, Edison Road, Basingstoke, Hants RG21 2YH, UK.
Tel (01256) 817282.
Fax (01256) 817292.

LSL **Life Sciences Laboratories**, 15 Ribocon Way, Progress Business Park (off Sedgewick Road), Luton, Beds LU4 9UR, UK.
Tel (01582) 597676.
Fax (01582) 581495.

FSL **Fisher Scientific UK**, Bishop Meadow Road, Loughborough, Leics LE11 5RG, UK.
Tel (01509) 231166.
Fax (01509) 231893.

HCL **Hichrom Ltd**, 1 The Markham Centre, Station Road, Theale, Reading, Berks RG7 4PE, UK.
Tel (01734) 303660.
Fax (01734) 323484.

ICN **ICN Biomedicals**, Thame Business Park, Wenman Road, Thame, Oxon OX9 3XA, UK.
Tel (01844) 213366, (0800) 282474.
Fax (01844) 213399, (0800) 614735.

ICN Pharmaceuticals Inc., Biomedical Products Division, 3300 Hyland Avenue, Costa Mesa, CA 92626, USA.
Tel (714) 545 0113, (800) 854 0530.
Fax (714) 557 4872, (800) 334 6999.

MBH **Merck Ltd**, Merck House, Poole, Dorset BH15 1TD, UK.
Tel (01202) 669700, (0800) 223344.
Fax (01202) 665599.

MCI **Mitsubishi Chemicals Industries Ltd**, Tokyo, Japan.
Tel (3) 3283 6254.
Fax (3) 3283 6287.

MNG • **Machery-Nagel GmbH**, Dueren, Germany.
Tel (2421) 9690.
Fax (2421) 969199.

MTC **Mitsui Toatsu Chemicals Inc.**, Chiyoda-Ku, Tokyo, Japan.
Tel (3) 3592 4111.

PEL **Perkin-Elmer Ltd**, Post Office Lane, Beaconsfield, Bucks HP9 2NE, UK.
Tel (01494) 676161.
Fax (01494) 679331/2/3.

Perkin-Elmer Corporation, 761 Main Avenue, Norwalk, CT 06859-0001, USA.
Tel (203) 761 5477, (800) 762 4000.
Fax (203) 761 5424.

PLC **Poly LC Inc.**, 9151 Rumsey Road, Suite 180, Columbia, MD 21045, USA.
Tel (410) 992 5400.
Fax (410) 730 8340.

PLL **Polymer Laboratories Ltd**, Essex Road, Church Stretton, Shropshire SY6 6AX, UK.
Tel (01694) 723581.
Fax (01694) 722171.

PMB **Pharmacia Biotech**, 23 Grosvenor Road, St Albans, Herts AL1 3AW, UK.
Tel (01727) 814000.
Fax (01727) 814001.
800 Centennial Avenue, PO Box 1327, Piscataway, NJ 08855-1327, USA.

PXG **Pentax GmbH**, Ulius Vosselierstrasse 104, 22527 Hamburg, Germany.
Tel (40) 561920.
Fax (40) 566475.

RIC **Rainin Instrument Co. Inc.**, Mack Road, Woburn, MA 01801, USA.
Tel (617) 935 3050.
Fax (617) 938 8157.

RTI **Regis Technologies Inc.**, 8210 Austin Avenue, PO Box 519, Morton Grove, IL 60053, USA.
Tel (847) 967 6000.
Fax (847) 967 5876.

SCC **Sigma-Aldrich Co. Ltd**, Fancy Road, Poole, Dorset BH12 4QH, UK.
Tel (01202) 733114, (0800) 373731.
Fax (01202) 715460, (0800) 378785.
Sigma Chemical Co., PO Box 14508, 3500 DeKalb Street, St Louis, MO 63178, USA.

Tel (908) 457 8000, (800) 526 3593.
Fax (908) 457 0557, (800) 329 3593.

PRC **Pierce**, 3747 North Meridian Road, PO Box 117, Rockford, IL 61105, USA.
Tel (815) 968 0747, (800) 874 3723.

PSB **PerSeptive Biosystems**, 3 Harforde Court, Foxholes Business Park, John Tate Road, Hertford, Herts SG13 7NW, UK.
Tel (01992) 507100.
Fax (01992) 553858.
500 Old Connecticut Path, Framingham, MA 01701, USA.
Tel (508) 383 7700, (800) 899 5858.
Fax (508) 383 7885.

PSL **Phase Separations Ltd**, Deeside Industrial Park, Deeside, Clwyd CH5 2NU, UK.
Tel (01244) 288500.
Fax (01244) 289500.

Tel (314) 771 5750, (800) 521 8956.
Fax (314) 771 5757, (800) 325 5052.

SCI **MICRA Scientific, Inc. (formerly SynChrom, Inc.)**, 1955 Techny Road, Suite 1, Northbrook, IL 60062, USA.
Tel (847) 272 7877.
Fax (847) 272 7893.

SFG **Boehringer Ingelheim Bioproducts Partnership (formerly Serva Feinbiochemica GmbH)**, Czernyring 22(11), 69115 Heidelberg, Germany.
Tel (6221) 598300.
Fax (6221) 598313.

SSP **Hypersil (formerly Shandon Southern Products Ltd)**, 112 Chadwick Road, Runcorn, Cheshire WA7 1PR, UK.
Tel (01928) 562633.
Fax (01928) 581078.

SUK **Supelco UK**, Fancy Road, Poole, Dorset BH12 4QH, UK.
Tel (01202) 716789, (0800) 887733.
Fax (01202) 715460, (0800) 378785.
Supelco Inc., Supelco Park, Bellefonte, PA 16833-0048, USA.
Tel (814) 359 3441.
Fax (814) 359 3044.

THS **TosoHass**, 156 Keystone Drive, Montgomeryville, PA 18936, USA.
Tel (215) 283 5000.
Fax (215) 283 5035.

TSG **The Separations Group (Vydac)**, PO Box 867, 17434 Mojave Street, Hesperia, CA 92345, USA.
Tel (760) 244 6107.
Fax (760) 244 1984.

VUL **Varian UK Ltd**, 28 Manor Rd, Walton-on-Thames, Surrey KT12 2QF, UK.
Tel (01932) 898000.
Fax (01932) 228769.

WIL **Whatman International Ltd**, Whatman House, St Leonard's Rd, 20/20 Maidstone, Kent ME16 0LS, UK.
Tel (01622) 676670.
Fax (01622) 677011.

REFERENCES

Chapter 1

1. Heftman, E. (1975) *Chromatography*. Van Nostrand Reinhold Co.
2. Gorbunoff, M.J. (1984) *Anal. Biochem.* **136**, 425.

Chapter 3

1. Hjerten, S. (1984) *Trends Anal. Chem.* **3**, 87.

Chapter 4

1. Harris, E.L.V. and Angal, S. (1989) *Protein Purification Methods: a Practical Approach*. IRL Press at Oxford University Press, Oxford.
2. Chaplin, M.F. and Kennedy, J.F. (1994) *Carbohydrate Analysis: a Practical Approach*, 2nd Edn. IRL Press at Oxford University Press, Oxford.
3. Converse, C.A. and Skinner, E.R. (1992) *Lipoprotein Analysis: a Practical Approach*. IRL Press at Oxford University Press, Oxford.
4. Lim, C.K. (1986) *HPLC of Small Molecules: a Practical Approach*. IRL Press at Oxford University Press, Oxford.
5. Marinetti, G. (1982) *Liquid Chromatographic Analysis*, Vol. 1. Marcel Dekker, New York.
6. Brown, P.R. (1984) *HPLC in Nucleic Acid Research*. Marcel Dekker, New York.
7. Oliver, R.W.A. (1989) *HPLC of Macromolecules: a Practical Approach*. IRL Press at Oxford University Press, Oxford.
8. Pazur, J.H. (1991) in *Advances in Carbohydrate Analysis*, Vol. 1 (C.A. White, ed.), p. 1. JAI Press, London.
9. Price, N.C. (1996) in *Proteins Labfax* (N.C. Price, ed.), p. 13. BIOS Scientific Publishers, Oxford.
10. Harris, E.L.V. (1989) in *Protein Purification Methods: a Practical Approach*. (E.L.V. Harris and S. Angal, eds), p. 125. IRL Press at Oxford University Press, Oxford.
11. Blin, N. and Stafford, D.W. (1976) *Nucl. Acids Res.* **3**, 2303.

12. Kirby, K.S. (1986) in *Methods Enzymol.* **XIIB**, 87.
13. Chomczynski, P. and Saachi, N. (1987) *Anal. Biochem.* **162**, 156.
14. Jones, P.G. *et al.* (1994) *RNA: Isolation and Analysis.* BIOS Scientific Publishers, Oxford.
15. Gallagher, J.T. (1989) in *HPLC of Macromolecules: a Practical Approach* (R.W.A. Oliver, ed.), p. 209. IRL Press at Oxford University Press, Oxford.
16. Tarentino, A.L. *et al.* (1985) *Biochemistry* **24**, 4664.
17. Tracey, B. (1986) in *HPLC of Small Molecules: a Practical Approach* (C.K. Lim, ed.), p. 69. IRL Press at Oxford University Press, Oxford.
18. Belfrage, P. and Vaughan, M. (1969) *J. Lipid Res.* **10**, 341.
19. Folch, J. *et al.* (1957) *J. Biol. Chem.* **226**, 497.
20. Slomiany, B.L. and Slomiany, A. (1977) *Biochim. Biophys. Acta* **486**, 531.
21. Andrews, A.G. (1984) *J. Chromatogr.* **336**, 139.
22. Lam, S. and Karmen, A. (1986) in *HPLC of Small Molecules: a Practical Approach* (C.K. Lim, ed.), p. 103. IRL Press at Oxford University Press, Oxford.
23. Niijima, S.-I. (1985) *Paediatr. Res.* **19**, 302.
24. Anderson, S.H.G. and Sjovall, J. (1984) *Anal. Biochem.* **134**, 309.
25. Shearer, M.J. (1986) in *HPLC of Small Molecules: a Practical Approach* (C.K. Lim, ed.), p. 157. IRL Press at Oxford University Press, Oxford.
26. Bieri, J.G. *et al.* (1985) *J. Liquid Chromatogr.* **8**, 473.
45. Ronin, C. *et al.* (1981) *Eur. J. Biochem.* **118**, 159.
46. Kivrikko, K.I. and Myllyla, R. (1980) in *The Enzymology of Post-translational Modification of Proteins*, Vol. 1 (R.B. Freedman and H.C. Hawkins, eds), p. 64. Academic Press, New York.
47. Maltese, W.A. (1990) *FASEB J.* **4**, 3319.
48. Rine, J.A. and Kim, S.-H. (1990) *New Biol.* **2**, 139.
49. Paik, W.K. and Kim, S. (1985) in *The Enzymology of Post-translational Modification of Proteins*, Vol. 2 (R.B. Freedman and H.C. Hawkins, eds), p. 187. Academic Press, New York.
50. Towler, D.A. *et al.* (1987) *Proc. Natl Acad. Sci. USA* **84**, 2708.
51. Towler, D.A. *et al.* (1987) *Proc. Natl Acad. Sci. USA* **84**, 2713.
52. Kennelly, P.J. and Krebs, E.G. (1991) *J. Biol. Chem.* **266**, 15555.
53. Casnellie, J.E. and Krebs, E.G. (1984) *Adv. Enzyme Regul.* **22**, 501.
54. Hortin, G.L. *et al.* (1986) *Biochem. Biophys. Res. Commun.* **141**, 326.
55. Riordan, J.F. and Vallee, B.L. (1972) *Methods Enzymol.* **25,** 494.
56. Boyd, H. *et al.* (1972) *Int. J. Peptide Protein Res.* **4**, 117.
57. Riordan, J.F. and Vallee, B.L. (1972) *Methods Enzymol.* **11**, 565.
58. Lindsay, D.G. and Shall, S. (1971) *Biochem. J.* **121**, 737.
59. Butler, P.J.G. and Hartley, B.S. (1972) *Methods Enzymol.* **25**, 191.
60. Klotz, I.M. (1967) *Methods Enzymol.* **11**, 570.
61. Hirs, C.H.W. (1967) *Methods Enzymol.* **11**, 548.
62. Yamada, H. *et al.* (1986) *J. Biochem.* **100**, 233.
63. Gurd, F.R.N. (1972) *Methods Enzymol.* **25**, 242.
64. Fields, R. (1972) *Methods Enzymol.* **25**, 464.

27. De Leenheer, A.P. *et al.* (1982) *J. Lipid Res.* **23**, 1362.
28. De Ruyter, M.G.M. and De Leenheer, A.P. (1978) *Clin. Chem.* **24**, 1920.
29. Gatautis, V.J. and Naito, H.K. (1981) *Clin. Chem.* **27**, 1672.
30. Wielders, J.P.M. and Mink, C.J.K. (1983) *J. Chromatogr.* **277**, 145.
31. Speek, A.J. *et al.* (1984) *J. Chromatogr.* **305**, 53.
32. Perret, D. (1986) in *HPLC of Small Molecules: a Practical Approach* (C.K. Lim, ed.), p. 221. IRL Press at Oxford University Press, Oxford.
33. Hart, D. and Piomelli, S. (1981) *Clin. Chem.* **27**, 220.
34. Garden, J.S. *et al.* (1977) *Clin. Chem.* **23**, 1585.
35. Lockwood, W.H. and Poulos, V. (1980) *Int. J. Biochem.* **12**, 1049.
36. Rossi, E. and Curnow, D.H. (1986) in *HPLC of Small Molecules: a Practical Approach* (C.K. Lim, ed.), p. 261. IRL Press at Oxford University Press, Oxford.
37. Scoble, H.A. *et al.* (1981) *Clin. Chim. Acta* **113**, 253.
38. Lee, F.-J.S. *et al.* (1990) *J. Biol. Chem.* **265**, 11576.
39. Rankin, P.W. *et al.* (1989) *J. Biol. Chem.* **264**, 4312.
40. Smets, L.A. *et al.* (1990) *Biochim. Biophys. Acta* **1054**, 49.
41. Bradbury, A.F. *et al.* (1982) *Nature* **298**, 686.
42. Price, P.A. *et al.* (1987) *Proc. Natl Acad. Sci. USA* **84**, 8335.
43. Cheung, A. *et al.* (1990) *Biochim. Biophys. Acta* **1039**, 90.
44. Gavel, Y. and von Heijne, G. (1990) *Protein Eng.* **3**, 433.
65. Ludwig, M.L. and Hunter, M.J. (1967) *Methods Enzymol.* **11**, 595.
66. Hunter, M.J. and Ludwig, M.L. (1972) *Methods Enzymol.* **25**, 585.
67. Inman, J.K. *et al.* (1983) *Methods Enzymol.* **91**, 559.
68. Stark, G.R. (1967) *Methods Enzymol.* **11**, 590.
69. Habeeb, A.F.S.A. (1972) *Methods Enzymol.* **25**, 558.
70. Kimmel, J.R. (1967) *Methods Enzymol.* **11**, 584.
71. Hounsell, E.F. (1986) in *HPLC of Small Molecules: a Practical Approach* (C.K. Lim, ed.), p. 49. IRL Press at Oxford University Press, Oxford.
72. Altmann, F. (1992) *Anal. Biochem.* **204**, 215.
73. Spiro, M.J. and Spiro, R.G. (1992) *Anal. Biochem.* **204**, 152.
74. Kang, E.Y.J. *et al.* (1990) *J. Protein Chem.* **9**, 31.
75. Seto, Y. and Shinohara, T. (1989) *J. Chromatogr.* **464**, 323.
76. El Rassi, Z. *et al.* (1991) *Carbohydr. Res.* **215**, 25.
77. Akiyama, T. (1991) *J. Chromatogr.* **588**, 53.
78. Webb, J.W. *et al.* (1988) *Anal. Biochem.* **169**, 337.
79. Kakehi, K. *et al.* (1991) *Anal. Biochem.* **199**, 256.
80. Takemoto, H. *et al.* (1985) *Anal. Biochem.* **145**, 245.
81. Alpenfels, W.F. (1981) *Anal. Biochem.* **114**, 153.
82. Hara, S. *et al.* (1989) *Anal. Biochem.* **179**, 162.
83. Christie, W.W. (1982) *J. Lipid Res.* **23**, 1072.
84. Christie, W.W. *et al.* (1984) *J. Chromatogr.* **298**, 513.
85. Lam, S. and Grushka, E. (1978) *J. Chromatogr.* **158**, 207.

Chapter 5

1. Jakoby, W.B. and Wilchek, M. (eds) (1974) *Methods Enzymol.* **XXXIV**.
2. Clonis, Y.D. (1989) in *HPLC of Macromolecules: a Practical Approach* (R.W.A. Oliver, ed.), p. 157. IRL Press at Oxford University Press, Oxford.
3. Dean, P.D.G. *et al.* (eds) (1985) *Affinity Chromatography, a Practical Approach.* IRL Press, Oxford.
4. Larsson, P.-O. (1984) *Methods Enzymol.* **104**, 212.
5. Regnier, F.E. and Nöel, R. (1976) *J. Chromatogr. Sci.* **14**, 316.
6. Kinkel, J.N. *et al.* (1984) *J. Chromatogr.* **297**, 167.
7. McGregor, J.L. *et al.* (1985) *J. Chromatogr.* **326**, 179.
8. Kita, K. *et al.* (1985) *Biochem. Int.* **10**, 319.
9. Montelaro, R.C. *et al.* (1981) *Anal. Biochem.* **114**, 398.
10. Tempst, P. *et al.* (1986) *J. Chromatogr.* **359**, 403.
11. Matson, R.S. and Goheen, S.C. (1986) *J. Chromatogr.* **359**, 285.
12. McKean, D.J. and Bell, M. (1982) *Protides Biol. Fluids* **30**, 709.
13. Winkler, G. *et al.* (1984) *J. Chromatogr.* **297**, 63.
14. Winkler, G. *et al.* (1985) *J. Chromatogr.* **326**, 113.
15. Welling, G.W. *et al.* (1984) *J. Chromatogr.* **297**, 101.
16. Calam, D.H. and Davidson, J. (1984) *J. Chromatogr.* **296**, 285.
17. Lambotte, P. *et al.* (1984) *J. Chromatogr.* **297**, 139.
18. Josic, D. *et al.* (1985) *Anal. Biochem.* **142**, 473.

37. Avrameas, S. and Ternynck, T. (1967) *Biochem. J.* **102**, 37C.
38. Zoller, M. and Matzku, S. (1976) *J. Immunol. Methods* **11**, 287.
39. Melchers, F. and Messer, W. (1970) *Eur. J. Biochem.* **17**, 267.
40. Weintraub, B.D. (1970) *Biochem. Biophys. Res. Commun.* **39**, 83.
41. Hill, R.J. (1972) *J. Immunol. Methods* **1**, 231.
42. Anderson, K.K. *et al.* (1979) *J. Immunol. Methods* **25**, 375.
43. Nicolson, G.L. and Blaustein, J. (1972) *Biochim. Biophys. Acta* **266**, 543.
44. Baues, R.J. and Gray, G.R. (1977) *J. Biol. Chem.* **252**, 57.
45. Olson, M.O.J. and Liener, I.E. (1967) *Biochemistry* **6**, 105.
46. Toyoshima, S. *et al.* (1970) *Biochim. Biophys. Acta* **221**, 514.
47. Entlicher, G. *et al.* (1970) *Biochim. Biophys. Acta* **221**, 272.
48. Vretblad, P. (1976) *Biochim. Biophys. Acta* **434**, 169.
49. Le Vine, D. *et al.* (1972) *Biochem. J.* **129**, 847.
50. Naemura, K. and Fukunaga, R. (1985) *Chem. Lett.* 1651.
51. Naemura, K. *et al.* (1985) *J. Chem. Soc. Chem. Commun.* 1560.
52. Kyba, E.P. *et al.* (1978) *J. Am. Chem. Soc.* **100**, 4555.
53. Peacock, S.C. *et al.* (1978) *J. Am. Chem. Soc.* **100**, 8190.
54. Peacock, S.C. *et al.* (1980) *J. Am. Chem. Soc.* **102**, 2043.
55. Lingenfelter, D.S. *et al.* (1981) *J. Org. Chem.* **46**, 393.
56. Yamamoto, K. *et al.* (1985) *J. Chem. Soc. Chem. Commun.* 1269.
57. Yamamoto, K. *et al.* (1986) *Bull. Chem. Soc. Jpn.* **59**, 1269.
58. Yamamoto, K. *et al.* (1987) *J. Chem. Soc. Chem. Commun.* 168.
59. Gehin, D. *et al.* (1986) *J. Org. Chem.* **51**, 1906.

19. Josic, D. *et al.* (1986). *J. Chromatogr.* **359**, 315.
20. Welling, G.W. *et al.* (1986) *J. Chromatogr.* **359**, 307.
21. Welling, G.W. *et al.* (1985) *J. Chromatogr.* **326**, 173.
22. Corran, P.H. (1989) in *HPLC of Macromolecules: a Practical Approach* (R.W.A. Oliver, ed.), p. 127. IRL Press at Oxford University Press, Oxford.
23. Plattner, R.D. (1981) *Methods Enzymol.* **72**, 21.
24. Kuksis, A. *et al.* (1983) *J. Chromatogr.* **273**, 43.
25. Marai, L. *et al.* (1983) *Can. J. Biochem. Cell Biol.* **61**, 840.
26. Herslof, B.G. and Pelura, T.J. (1982) *J. Am. Oil Chem. Soc.* **59**, 308A Abstr. No. 295.
27. El Hamdy, A.H. and Perkins, E.G. (1981) *J. Am. Oil Chem. Soc.* **58**, 867.
28. Shaw, R. *et al.* (1978) *Anal. Biochem.* **86**, 450.
29. Block, C. and Watkins, J.B. (1978) *J. Lipid Res.* **19**, 510.
30. Nakayama, F. and Nakagaki, M. (1980) *J. Chromatogr.* **183**, 287.
31. Wildgrude, H.J. *et al.* (1983) *J. Chromatogr.* **282**, 603.
32. Wildgrude, H.J. *et al.* (1986) *J. Chromatogr.* **353**, 207.
33. Hudson, L. and Hay, F.C. (1980) *Practical Immunology.* Blackwell Scientific Publications, Oxford.
34. Kristianson, T. (1978) in *Affinity Chromatography* (O. Hoffman-Ostenhoff *et al.*, eds). Pergamon, New York.
35. Chidlow, J.W. *et al.* (1974) *FEBS Lett.* **41**, 248.
36. Mains, R.E. and Eipper, B.A. (1976) *J. Biol. Chem.* **251**, 4115.
60. Chadwick, J.D. *et al.* (1984) *J. Chem. Soc. Perkin Trans.* **I**, 1707.
61. Davidson, R.B. *et al.* (1984) *J. Org. Chem.* **49**, 353.
62. Nakazaki, M. *et al.* (1983) *J. Chem. Soc. Chem. Commun.* 787.
63. Naemura, K. *et al.* (1985) *J. Chem. Soc. Chem. Commun.* 1560.
64. Thoma, A.P. *et al.* (1979) *Membren. Helv. Chim. Acta* **62**, 2303.
65. Behr, J.-P. *et al.* (1980) *Helv. Chim. Acta* **63**, 2096.
66. Krstulovic, A.M. (1989) *Chiral Separations by HPLC.* Ellis Horwood, Chichester.
67. Hare, P.E. and Gil-Av, E. (1979) *Science* **104**, 1226.
68. Gil-Av, E. *et al.* (1980) *J. Am. Chem. Soc.* **102**, 5115.
69. Oelrich, E. *et al.* (1980) *J. High Resolut. Chromatogr. Chromatogr. Commun.* **3**, 269.
70. Gilon, C. *et al.* (1979) *J. Am. Chem. Soc.* **101**, 7612.
71. Gilon, C. *et al.* (1980) *Anal. Chem.* **52**, 1206.
72. Gilon, C. *et al.* (1981) *J. Chromatogr.* **203**, 365.
73. Grushka, E. *et al.* (1983) *J. Chromatogr.* **255**, 41.
74. Nimura, N. *et al.* (1981) *Anal. Chem.* **53**, 1380.
75. Nimura, N. *et al.* (1982) *J. Chromatogr.* **234**, 482.
76. Nimura, N. *et al.* (1982) *J. Chromatogr.* **239**, 671.
77. Nimura, N. *et al.* (1984) *J. Chromatogr.* **316**, 547.
78. Weinstein, S. (1982) *Angew. Chem. Suppl.* 425.
79. Weinstein, S. *et al.* (1982) *Anal. Biochem.* **121**, 370.
80. Weinstein, S. and Weiner, S. (1984) *J. Chromatogr.* **303**, 244.
81. Weinstein, S. and Grinberg, N. (1985) *J. Chromatogr.* **318**, 117.

82. Kurganov, A. and Davankov, V. (1981) *J. Chromatogr.* **218**, 559.
83. Wernicke, R. (1985) *J. Chromatogr. Sci.* **23**, 39.
84. LePage, J. *et al.* (1979) *Anal. Chem.* **51**, 433.
85. Lindner, W. *et al.* (1979) *J. Chromatogr.* **185**, 323.
86. Tapuhi, Y. *et al.* (1981). *J. Chromatogr.* **205**, 325.
87. Lam, S. and Chow, F. (1980) *J. Liquid Chromatogr.* **3**, 1579.
88. Lam, S. *et al.* (1980) *J. Chromatogr.* **199**, 295.
89. Lam, S. (1982) *J. Chromatogr.* **234**, 483.
90. Lam, S. and Karmen, A. (1982) *J. Chromatogr.* **239**, 451.
91. Lam, S. and Karmen, A. (1984). *J. Chromatogr.* **289**, 339.
92. Lam, S. (1984) *J. Chromatogr. Sci.* **22**, 416.
93. Klemisch, W. *et al.* (1981) *J. High Resolut. Chromatogr. Chromatogr. Commun.* **4**, 535.
94. Horikawa, R. *et al.* (1986) *J. Liquid Chromatogr.* **9**, 537.
95. Lindner, W.F. and Hirschbock, I. (1986) *J. Liquid Chromatogr.* **9**, 551.
96. Forsman, U. (1984) *J. Chromatogr.* **303**, 217.
97. Mefford, I.N. and Barchas, J.D. (1980) *J. Chromatogr.* **181**, 187.
98. Reinhard, J.F. Jr. *et al.* (1980) *Life Sci.* **27**, 905.
99. Mefford, I.N. (1981) *J. Neurosci. Methods* **3**, 207.
100. Matson, W.R. *et al.* (1984) *Clin. Chem.* **30**, 1477.
101. Langlais, P.J. *et al.* (1984) *Clin. Chem.* **30**, 1046.
102. Saller, C.F. and Salama, A.I. (1984) *J. Chromatogr.* **309**, 287.
103. Bieri, J.G. *et al.* (1985) *J. Liquid Chromatogr.* **8**, 473.

9. Derler, H. *et al.* (1988) *J. Chromatogr.* **440**, 281.
10. Hicks, K.B. and Sondey, S.M. (1987) *J. Chromatogr.* **389**, 183.
11. Satyaswaroop, P.G. *et al.* (1977) *Steroids* **30**, 139.
12. Hernandez, L.M. *et al.* (1990) *Carbohydr. Res.* **203**, 1.
13. Reddy, G.P. and Bush, A.C. (1991) *Anal. Biochem.* **198**, 278.
14. Zopf, D. *et al.* (1989) *Methods Enzymol.* **179**, 55.
15. Wang, W. *et al.* (1992) *Arch. Biochem. Biophys.* **292**, 433.
16. Ammeraal, R.N. *et al.* (1991) *Carbohydr. Res.* **215**, 179.
17. Townsend, R.R. et al. (1988) *Methods Enzymol.* **179**, 65.
18. Watanabe, K. (1985) *J. Chromatogr.* **337**, 126.
19. Hall, N.A. and Patrick, A.D. (1989) *Anal. Biochem.* **178**, 378.
20. Kennedy, J.F. *et al.* (1985) *Food Chem.* **16**, 115.
21 Kennedy, J.F. *et al.* (1985) *Food Chem.* **18**, 95.
22. Cumming, D.A. *et al.* (1988) *Carbohydr. Res.* **179**, 369.
23. Hoffmann, R.A. *et al.* (1991) *Carbohydr. Res.* **221**, 63.
24. Gruppen, H. *et al.* (1992) *Carbohydr. Res.* **233**, 45.
25. Goux, W.J. (1988) *Carbohydr. Res.* **173**, 292.
26. Ito, Y. *et al.* (1987) *J. Chromatogr.* **391**, 296.
27. Perrett, D. (1986) in *HPLC of Small Molecules: a Practical Approach* (C.K. Lim, ed.), p. 221. IRL Press at Oxford University Press, Oxford.
28. Bennett, G.W. *et al.* (1981) *Life Sci.* **29**, 1001.
29. Borch, R.F., *et al.* (1975) *Anal. Chem.* **47**, 2437.
30. Durst, H.D., *et al.* (1975) *Anal. Chem.* **47**, 1797.

104. De Leenheer, A.P. *et al.* (1982) *J. Lipid Res.* **23**, 1362.
105. De Ruyter, M.G.M. and De Leenheer, A.P. (1978). *Clin. Chem.* **24**, 1920.
106. Gatautis, V.J. and Naito, H.K. (1981) *Clin. Chem.* **27**, 1672.
107. Wielders, J.P.M. and Mink, C.J.K. (1983) *J. Chromatogr.* **277**, 145.
108. Speek, A.J. *et al.* (1984) *J. Chromatogr.* **305**, 53.
109. O'Riordan, J.L.H. *et al.* (1982) in *Vitamin D: Chemical, Biochemical and Clinical Endocrinology of Calcium Metabolism* (A.W. Norman *et al.*, eds), p. 751. Walter de Gruyter, Berlin.

Chapter 6

1. Scott, R.P.W. (1986) *Liquid Chromatography Detectors*, 2nd Edn. Elsevier, Amsterdam.
2. Ben-Bassat, A.A. and Grushka, E. (1991) *J. Liquid Chromatogr.* **14**, 1051.
3. De Ruyter, M.G.M. and De Leenheer, A.P. (1978) *Clin. Chem.* **24**, 1920.
4. De Leenheer, A.P. *et al.* (1982) *J. Lipid Res.* **23**, 1362.
5. Bieri. J.G. *et al.* (1985) *J. Liquid Chromatogr.* **8**, 473.
6. Seki, T. and Yamaguchi, Y. (1984) *J. Chromatogr.* **305**, 188.
7. Koizumi, K. *et al.* (1987) *J. Chromatogr.* **409**, 396.
8. Scott, F.W. and Hatina, G. (1988) *J. Food Sci.* **53**, 264.
31. Grushka, E. *et al.* (1975) *J. Chromatogr.* **112**, 673.
32. Fitzpatrick, F.A. (1976) *Anal. Chem.* **48**, 499.
33. Ahmed, M.S. *et al.* (1980) *J. Chromatogr.* **192**, 387.
34. Nagels, L. *et al.* (1980) *J. Chromatogr.* **190**, 411.
35. Pierce Technical Bulletin: Preparation of Phenacyl and *p*-Bromophenacyl Derivatives for HPLC.
36. Knapp, D. and Krueger, S. (1975) *Anal. Lett.* **8**, 603.
37. Pierce Technical Bulletin: Preparation of *p*-Nitrobenzyl Derivatives for HPLC.
38. Chang, J.Y. (1981) *Biochem. J.* **199**, 537.
39. Chang, J.Y. (1981) *Biochem. J.* **199**, 557.
40. Chang, J.Y. *et al.* (1982) *Eur. J. Biochem.* **127**, 625.
41. Chang, J.Y. (1983) *Methods Enzymol.* **91**, 455.
42. Chang, J.Y. *et al.* (1983) *Biochem. J.* **211**, 173.
43. Knecht, R. *et al.* (1983) *Anal. Biochem.* **130**, 65.
44. Foriers, A. *et al.* (1984) *J. Chromatogr.* **297**, 75.
45. Yang, C.Y. and Wakil, S.J. (1984) *Anal. Biochem.* **137**, 54.
46. Stocchi, V. *et al.* (1985) *J. Chromatogr.* **349**, 77.
47. Lin, J.K. *et al.* (1980) *Anal. Chem.* **52**, 630.
48. Chang, J.Y. *et al.* (1981) *Biochem. J.* **199**, 547.
49. Chang, J.Y. *et al.* (1981) *FEBS Lett.* **132**, 117.
50. Chang, J.Y. *et al.* (1982) *Biochem. J.* **203**, 803.
51. Chang, J.Y. (1984) *J. Chromatogr.* **295**, 193.
52. Vendrell, J. and Aviles, F.X. (1986) *J. Chromatogr.* **358**, 401.

53. Lindner, W. *et al.* (1979) *J. Chromatogr.* **185**, 323.
54. Tapuhi, Y. *et al.* (1981) *Anal. Biochem.* **115**, 123.
55. Tapuhi, Y. *et al.* (1981) *J. Chromatogr.* **205**, 325.
56. Casoli, A. and Colugrande, O. (1982) *Amer. J. Enol. Vitic.* **33**, 135.
57. Grego, B. and Hearn, M.T.W. (1983) *J. Chromatogr.* **255**, 67.
58. Oray, B. *et al.* (1983) *J. Chromatogr.* **270**, 253.
59. Miyano, H. *et al.* (1985) *Anal. Biochem.* **150**, 125.
60. Marfey, P. (1984) *Carlsberg Res. Commun.* **49**, 591.
61. Aberhart, D.J. *et al.* (1985) *Anal. Biochem.* **151**, 88.
62. Szokan, G. *et al.* (1988) *J. Chromatogr.* **444**, 115.
63. Böhlen, P. *et al.* (1979) *Anal. Biochem.* **94**, 313.
64. Jones, B.N. and Gilligan, J.P. (1983) *Am. Biotech. Lab.* **12**, 46.
65. Jones, B.N. and Gilligan, J.P. (1983) *J. Chromatogr.* **266**, 471.
66. Seaver, S.S. (Sept/Oct 1984) *Biotechniques* 254.
67. Fried, V.A. *et al.* (1985) *Anal. Biochem.* **146**, 271.
68. Fiedler, H.P. *et al.* (1986) *J. Chromatogr.* **353**, 201.
69. Stein, W.H. and Moore, S. (1949) *Cold Spring Harbor Symp. Quant. Biol.* **14**, 179.
70. Heinrikson, R.L. and Meredith, S.C. (1984) *Anal. Biochem.* **136**, 65.
71. Scholze, H. (1985) *J. Chromatogr.* **350**, 453.
72. Caudill, W.L. *et al.* (1982) *J. Chromatogr.* **227**, 331.
73. Caudill, W.L. *et al.* (1982) *Bioanal. Syst. Curr. Separ.* **4**, 59.
74. Andrews, J.L. *et al.* (1982) *Arch. Biochem. Biophys.* **214**, 386.
75. White, C.A. and Kennedy, J.F. (1981) *Tech. Life Sci.* **B3**, B312/1.
76. James, L.B. (1984) *J. Chromatogr.* **284**, 97.
77. Perini, F. and Peters, B.P. (1982) *Anal. Biochem.* **123**, 357.
78. Sinner, M. and Puls, J. (1978) *J. Chromatogr.* **156**, 197.
79. Honda, S. and Suzuki, S. (1984) *Anal. Biochem.* **142**, 167.
80. Bach, E. and Schollmeyer, E. (1992) *Anal. Biochem.* **203**, 335.
81. Honda, S. and Suzuki, S. (1984) in *Proceedings of the XIIth International Carbohydrate Symposium* (J.F.G. Vliegenthart *et al.*, eds), p. 501. Vonk Publishers, Zeist.
82. Kato, T. and Kinoshita, T. (1980) *Anal. Biochem.* **106**, 238.
83. Grimble, G.K. *et al.* (1983) *Anal. Biochem.* **128**, 422.
84. Tomiya, N. *et al.* (1992) *Anal. Biochem.* **206**, 98.
85. Peelen, G.O.H. *et al.* (1991) *Anal. Biochem.* **198**, 334.
86. D'Amboise, M. et al. (1980) *Carbohydr. Res.* **79**, 1.
87. Manzi, A.E. *et al.* (1990) *Anal. Biochem.* **188**, 20.
88. Pande, S.V. (1976) *Anal. Biochem.* **74**, 25.
89. Fricke, U. (1975) *Anal. Biochem.* **63**, 555.

INDEX